# Coarse Fishing
# Basics

# Coarse Fishing Basics

## Steve Partner

First published in Great Britain in 2006 by

Hamlyn, a division of Octopus Publishing Group Ltd

Hardback edition published in 2007 by Bounty Books, a division of Octopus Publishing Group Ltd

Reprinted 2008

This paperback edition published in 2013 by Bounty Books, a division of Octopus Publishing Group Ltd

Endeavour House, 189 Shaftesbury Avenue, London WC2H 8JY

www.octopusbooks.co.uk

An Hachette UK Company

www.hachette.co.uk

Copyright © Octopus Publishing Group Ltd 2006

ISBN: 978-0-753725-19-1

A CIP catalogue record for this book is available from the British Library

Printed and bound in China

# contents

| | |
|---|---|
| introduction | 6 |
| getting started | 8 |
| rods | 10 |
| reels | 14 |
| poles | 18 |
| tackle sundries | 20 |
| how do I ... ? | 26 |
| ... spool a reel | 28 |
| ... learn to cast | 30 |
| ... shot a float | 32 |
| ... set up a leger | 34 |
| ... use a pole | 36 |
| ... mix groundbait | 38 |
| ... tie basic knots | 40 |
| which bait? | 42 |
| bread | 44 |
| boilies | 46 |
| maggots and casters | 48 |
| worms and other naturals | 50 |
| seeds, pulses and nuts | 52 |
| pellets, paste and luncheon meat | 54 |
| deadbaits and lures | 56 |
| know your fish | 58 |
| barbel | 60 |
| bream | 62 |
| wels catfish | 64 |
| dace | 65 |
| common carp | 66 |
| chub | 68 |
| rudd | 70 |

| | |
|---|---|
| crucian carp | 71 |
| perch | 72 |
| pike | 74 |
| roach | 76 |
| eel | 78 |
| zander | 79 |
| tench | 80 |
| other common species | 82 |
| finding fish | 84 |
| in lakes and ponds | 86 |
| in canals | 88 |
| in rivers | 90 |
| in weir pools | 92 |
| in gravel pits | 94 |
| essential skills | 96 |
| choosing a swim | 98 |
| playing and landing fish | 100 |
| unhooking your catch | 102 |
| legering for bream | 104 |
| bolt-rigging for carp | 106 |
| trotting for roach | 108 |
| deadbaiting for pike | 110 |
| stalking for chub | 112 |
| floodwater fishing for barbel | 114 |
| lure fishing for perch | 116 |
| using the lift method for tench | 118 |
| 50 best fishing tips | 120 |
| fishing in Europe | 124 |
| index | 126 |
| acknowledgements | 128 |

# Introduction

**Left** *An angler using a pole takes the strain as a good fish makes a bid for freedom.*

## jargon buster

**Coarse fish** Includes all freshwater fish apart from members of the salmon family.

**Seafish** Saltwater species that can be caught only at sea or from the shore.

**Game fish** Freshwater fish of the salmon family, including trout.

**Match anglers** Anglers who compete against each other over a set period, with the winner being the one who catches the greatest overall weight.

Angling is one of Europe's most popular pastimes, and every weekend, all across the continent, millions of people pick up a rod or pole and pit their wits against an underwater foe in the name of sport.

The word angling embraces three broad disciplines – sea, game and coarse fishing – all of which have their own tackle, methods and rules and regulations. Sea anglers concentrate on catching fish from the oceans and usually eat what they catch. Game anglers tend to use artificial flies to tempt species like trout and salmon from rivers, streams and purpose-built reservoirs. Most will eat what they catch.

Coarse fishing, which is by far the most popular of the three types, is done with a range of light tackle on inland lakes, canals and rivers for fish such as roach, perch, bream, carp and pike. In Britain coarse fishing is regarded as a sport, and the fish are caught for the challenge and pleasure of it and are safely returned to the water after being caught. In some other parts of Europe, however, coarse fish are destined for the table.

## So much choice

What makes coarse fishing so popular is the sheer breadth of choice. Nearly everywhere you find water you'll also find fish, and with a little knowledge and skill you will be able to catch the fish of your choice in a wide range of conditions. Some anglers target carp and pike, chasing monsters that weigh more than 18kg (40lb), while match anglers compete against each other in order to see who can catch the greatest weight of fish in an allotted time from a particular stretch of water.

## What you'll learn

Of course, as with every sport, becoming a good angler will take time and patience, and your first step is to familiarize yourself with the basics – and these you'll find contained within the pages of this book.

You'll learn about tackle, baits, the different types of fish and where to find them, as well as a host of simple, yet effective, methods that will help you catch them. Don't be in a rush to get out to the lakeside without first arming yourself with the minimum of knowledge, because an ill-prepared angler is likely to fail. Instead, get to understand the basics and be patient and sensible, and the chances are that you'll catch fish – and that, in turn, will transform you into an angler for life.

Fishing is constantly changing, and the best anglers are those who are willing to learn, no matter how many years of experience they may have.

## Get out there – and enjoy it!

Books, however, will never beat being outside and actually fishing. If you can find more experienced anglers to point you in the right direction, this will help enormously and you'll certainly find that anglers are among the most encouraging of people, who are usually more than happy to share information and their experiences with you.

Remember, though, that coarse fishing isn't just about catching fish. It's about appreciating the countryside and the wildlife you'll find in it. It's about the joy of pitting your wits against a wild, unseen foe. It's about the camaraderie and fun between fellow anglers. But most of all it's about the simple enjoyment of being beside the water, and catching a fish is often merely a bonus to a thoroughly rewarding outdoor experience. So get out there and enjoy being an angler!

**Above** *You'll only catch fish if you can attract them to your swim and a bait-dropper is a great way of doing it.*

**Above** *With the rods cast out, it can be a waiting game for the fish to arrive.*

# getting
## started

# getting started
# rods

## jargon buster

**Carbonfibre** A strong, lightweight material used in the manufacture of fishing rods.

**Fibreglass** A material widely used to make fishing rods from the 1950s to the 1980s.

**Action** The way in which a rod bends.

**Float** A small piece of buoyant plastic, quill or balsa wood used to present a bait.

**Legering** Fishing with the bait on the bottom, where it is kept by means of a weight.

**Swimfeeder** A perforated plastic tube from which bait can escape.

**Leger weight** A small piece of lead that allows the angler to fish on the bottom.

**Quivertip** A slim length of carbonfibre or fibreglass that acts as a bite indicator.

**Test curve** The amount of weight needed to make the rod bend to its greatest extent.

**Spinning** The art of retrieving a revolving lure with a hinged, shiny blade that flashes underwater and carries a single treble hook.

Fishing rods have come a long way since the time when a piece of garden cane or even a discarded aerial were used. Nowadays the choice of rods is bewildering, but beginners really need only one or two different kinds to cover most types of angling. A general float rod and one for legering will cope with almost any scenario that you will initially encounter.

## the anatomy of a rod

**tip section**
The tip section is the end of the rod, which is much thinner than the rest, and its design determines the sort of action the rod has – that is, the way it bends. Tip-action rods bend only at the tip and are for fast-striking and small fish; middle-action rods bend from the tip to the centre and are for bigger fish; and through-action rods bend all the way to the butt and are designed for big fish at close range.

**rings**
All rods feature a series of rings that are whipped on to the blank with cotton and are then glued and varnished to keep them in place. The number of rings depends on the length of the rod and what it's designed for. Float rods have lots of rings; leger rods have fewer.

**blank**
The word used to describe the basic rod before the handle, reel seat and rings are added.

**butt section**
Rods are generally made in two or three sections, and the butt section is the one that features the handle.

**reel seat**
These can be screw-in or sliding. Screw-in seats tend to be more secure and are usual on modern rods; sliding seats are more likely to be found on older models.

**handle**
The handle of the standard fishing rod used to be made of cork; in more recent times a material known as duplon has become popular.

**Above** *The rod bends over as the angler strikes into a fish. Choosing the right rod for the job is crucial.*

Most modern rods are made from carbonfibre, which is both light and flexible, although cheaper versions are constructed from fibreglass, which tends to make them heavier. Rods can defined by their action (see page 10). A stiff action is usually better suited to bigger fish, while a softer one is preferable for smaller quarry.

There are six main types of fishing rod that cater for almost all the scenarios you are likely to encounter: float rod, leger rod, specialist rod, spinning rod, carp rod and pike rod. The key to success is making sure that you use the right tool for the right job – there's no point attempting to target big fish with a float rod, for example.

**Float rod** *Probably the most common and versatile of all rods, these are designed for use with a float.*

## float rod

The most common type of rod and the one that most beginners will start with is a float rod. Designed, as the name suggests, for use with floats, this sort of rod usually has three sections and is about 4m (13ft) long. The rod is this length to give the angler better control of the float.

The action of most float rods is all-through, meaning that when they are bent they arc all the way down the rod to the butt. This action suits a light line and small hooks, which are used when the target is small fish.

Most float rods have cork handles and fixed reel seats so that the reel sits firmly on the rod handle and cannot become detached while you are in the process of fishing.

## leger rod

Leger rods are designed for use with leger weights or swimfeeders (see pages 34–5). They are a little stronger than float rods and are also shorter, usually averaging about 3.5m (11ft 6in).

The most popular style of leger rod for all-round fishing is called a quivertip rod. These come with a built-in quivertip – a sensitive, brightly coloured tip at the end of the rod. The angler uses this tip to detect bites. Rods like this also have a through action and are designed for relatively light lines and small to average-sized fish.

**Leger rod** *Designed for use with leger weights and swimfeeders, anglers detect a bite through the use of quivertips.*

*Specialist rod These tend to be made of two sections and are stronger to deal with bigger fish.*

**Above** *Balanced tackle is key, so choose the right rod for the right type of fishing.*

## specialist rod

Specialist rods are the preserve of the angler who wishes to target big fish, such as chub, barbel and tench. These rods pack plenty of power and typically have a test curve of 0.68–0.9kg (1lb 8oz–2lb). A test curve is determined by the weight needed to make a rod bend to its maximum, and the larger the fish you will be targeting, the higher the test curve should be.

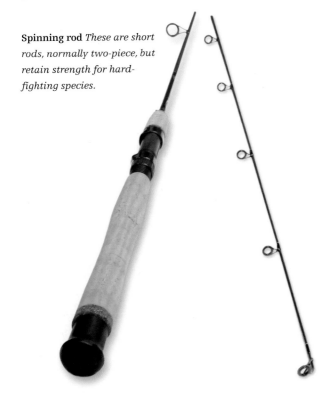

**Spinning rod** *These are short rods, normally two-piece, but retain strength for hard-fighting species.*

## spinning rod

These are probably the shortest rods on the market, with an average length of 2.4–2.7m (8–9ft). The key to a good spinning rod is weight, because you can spend hours casting and retrieving a lure. Designed for hard-fighting predators like pike, these rods are built with plenty of strength.

## carp rod

Carp rods need to be really robust, and some have test curves of 1.36kg (3lb), although most have a test curve of 1.13kg (2lb 8oz). This power is necessary not only because you will be casting heavy leger weights but may also be playing fish that weigh up to 20kg (44lb)! Carp rods are made in two pieces and have fewer rings than float and leger rods. They are usually about 3.6m (12ft) long.

## pike rod

Like carp rods, pike rods are heavy-duty tools, designed for casting big baits long distances, but they must also be flexible enough to prevent hook pulls from wary pike that may not have got hold of the bait properly.

If you will be pike fishing only occasionally, a carp rod will suffice, but if you become serious about pike, invest in one specifically designed for the job. These rods come in two sections and can boast a test curve of up to 1.58kg (3lb 8oz).

# getting started
# reels

**Left** *Without a reel a rod is useless. Choose the right type of reel for the type of fishing you want to do.*

## jargon buster

**Spool** A cylinder that is pushed on to the body of the reel and that revolves when the handle is turned. Line is stored on it.

**Bale arm** A metal arm on a fixed-spool reel that keeps the line from peeling off the spool and becoming tangled.

**Drag** An adjustable knob on the back of the reel body, which can be adjusted to allow more or less line to be pulled off the spool by a fish.

**Trotting** Using a float on a river in such a way that the bait is presented naturally, with the flow of the water downstream.

**Anti-reverse lever** A switch, often under the reel body, that, when disengaged, enables the handle to spin backwards, allowing a fish to take line.

**Freespool** A function that allows line to be removed from the reel without releasing the bale arm.

**Line clip** Found on the spool, this is used for trapping the main line when accuracy is needed. It is frequently used by anglers who are legering for bream.

The fishing reel is an essential piece of kit if you want to floatfish or leger with a rod. It enables you to retrieve line and then cast out again. There are four basic designs: the fixed-spool reel, which is for general use; the closed-face reel, which is for match and river fishing; the freespool reel, which is for big fish; and the centrepin, which is the preferred reel for river specialists.

For beginners, a simple fixed-spool reel is the best place to start, with the other types built for more specialist angling. Like every other item of fishing tackle you will need, always buy the best you can afford – cutting corners is false economy.

## the anatomy of a reel

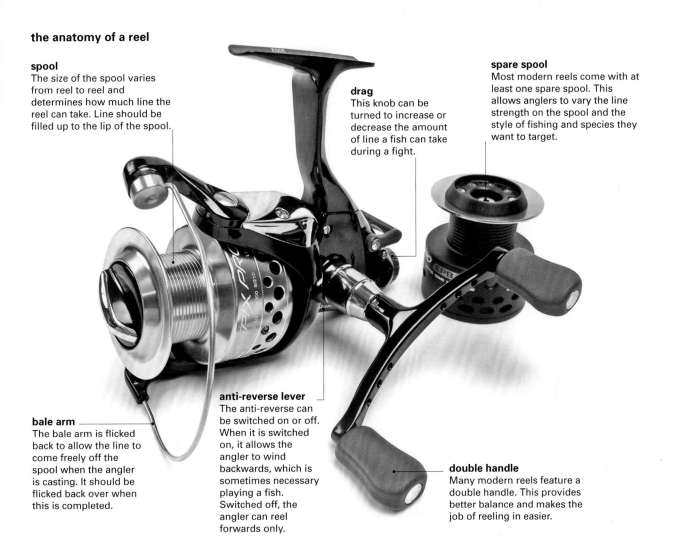

**spool**
The size of the spool varies from reel to reel and determines how much line the reel can take. Line should be filled up to the lip of the spool.

**drag**
This knob can be turned to increase or decrease the amount of line a fish can take during a fight.

**spare spool**
Most modern reels come with at least one spare spool. This allows anglers to vary the line strength on the spool and the style of fishing and species they want to target.

**bale arm**
The bale arm is flicked back to allow the line to come freely off the spool when the angler is casting. It should be flicked back over when this is completed.

**anti-reverse lever**
The anti-reverse can be switched on or off. When it is switched on, it allows the angler to wind backwards, which is sometimes necessary playing a fish. Switched off, the angler can reel forwards only.

**double handle**
Many modern reels feature a double handle. This provides better balance and makes the job of reeling in easier.

The basic fixed-spool reel hasn't changed very much since it was first invented in the early 1900s. The basic design is much the same, although the materials from which they are now made are lighter, stronger and more durable than ever before.

All fixed-spool reels – which are by far the most popular of the several different types of reel available – feature a series of characteristics that enable them to do their job properly.

They all have a bale arm, which keeps the line from peeling from the spool and causing tangles, a handle for retrieving the line, a rear drag that allows line to be taken by a fighting fish and an anti-reverse lever. This allows the handle to wind backwards and is used as an alternative for releasing line when playing a big fish.

## How do reels work?

All reels – whether they be fixed-spool reels or otherwise – work on a very simple principle. They allow the angler to cast his rig into the water and then retrieve it at a later period. When the handle is turned, the spool rotates and gathers line in. Then, when the bale arm is opened and the rig cast, line is released.

Even the most expensive reels are prone to tangling when in the hands of the inexperienced. Don't panic if this happens – there isn't an angler in the world who hasn't experienced a 'bird's nest' (so called because of the resemblance a ball of tangled line has to a bird's nest) at some time or another!

One way to minimize this happening is to correctly spool the line on to your reel – a skill you will later learn in this book.

**fixed-spool reel**                                    **closed-face reel**

## fixed-spool reels

These are the most common reels on the market,
and they are the easiest to operate. They consist of
a strong wire arm, known as the bale arm
(sometimes spelled bail arm), which can be folded
back across the spool to free the line from the
spool for casting. Once the cast is complete, a
simple turn of the reel handle re-engages the arm,
bringing it back into position and trapping the
line back on the spool.

Fixed-spool reels usually have a small lever,
normally situated beneath the reel, which is used
when a big fish is hooked. This is called an anti-
reverse lever, and it allows the spool to rotate and
release line without opening the bale arm.

This type of reel also has an adjustable 'drag',
found at the back of the reel, which performs a
similar function but can be tightened or loosened
to determine when line is released.

Look out for the line clip, which will be found
on the spool. It is used to trap line when you are
fishing with a swimfeeder or leger weight (see
pages 34–5).

## closed-face reels

One of the main drawbacks with fixed-spool reels
is that the spool is open to the elements,
especially the wind, which can create tangles. The
closed-face reel overcomes this problem. Basically,
the spool's bale arm is replaced with a removable
metal cover from which the line exits via a slot.
The line is held in place by a pin that retracts
inside the spool when a button on the front of the
reel is pressed. This allows the angler to cast.

A problem with closed-face reels is that there
is no drag facility, which makes giving line to
running fish difficult. For this reason, closed-face
reels are most often used for small fish, especially
for trotting on rivers (see pages 108–9).

They are a favourite with match anglers,
especially those that fish on running water,
because of the extra control of the float that they
allow. However, they are probably best avoided
initially by the beginner in favour of a more
simple fixed-spool reel.

Closed face reels tend to be more expensive too
and are not as readily available.

**centrepin reel**

**freespool reel**

## centrepin reels

These reels are not designed for beginners, and they require a considerable amount of skill to operate. Designed almost exclusively for use on running water, centrepin reels have two rotating metal discs, with the line held on a spool between the discs. Line is removed from the spool by the pace of the flow on a river, and in the right hands they are extremely effective.

The main difficulty with a centrepin reel lies in casting. It is not simply a case of removing a bale arm and allowing line to be freely released. Instead, the angler must draw line from each of the rings before making an exaggerated sweep of the arm.

Centrepin reels are more commonly found among the game fishing fraternity, where they lend themselves to casting imitation flies for species like trout and salmon.

Those that use them in coarse fishing are almost exclusively river anglers who enjoy the way that they allow a float to travel 'naturally' with the flow. They are also more expensive than simple fixed-spool reels.

## freespool reels

This type of reel was designed in the 1970s by carp and big-fish anglers. A freespool reel is a normal fixed-spool reel with a special facility that allows line to be released without opening the bale arm or fiddling with the drag.

When a fish picks up the bait and moves off, the line peels effortlessly from the spool. But once the rod is picked up, a simple turn on the reel disengages the freespool facility so that it reverts to behaving like a normal fixed-spool reel.

Freespool reels are commonly used by anglers legering for carp, pike, bream, barbel and tench.

The obvious difference you will notice about this type of reel is the size. They are often considerably bigger than fixed-spool reels and a fair deal heavier too. This is because the spool is larger to allow for heavier and thicker line needed to tackle bigger-than-average fish.

These are very much specialist tools and it is unlikely that the beginner will need to encounter them until they are more experienced. Expect to pay more for a freespool reel too.

# getting started
# poles

## jargon buster

**Margins** The area of a swim nearest to the angler. This will often be right at the end of the rod tip.

**Snags** Immovable features in the swim, such as trees, sunken obstacles and reeds, which provide ambush points for predators.

**Flicktip** The end of a whip. It is made from flexible carbonfibre, which acts as a shock absorber. It is used only when small fish are the target.

Pole fishing might appear to be the simplest form of angling: a line is tied to one end of a straight piece of material. These days, however, it's a lot more complicated than that.

Like rods, poles are made from a number of materials, but carbonfibre is the best, being light and strong. Some are made from a blend of fibreglass and carbonfibre, and these tend to be cheaper.

Poles are generally much longer than rods – some are up to 16m (more than 52ft) long – although most tend to be about 11m (36ft). They are made up of separate sections, usually about 1m (3ft) each, so that a 10m (33ft) pole can be made of about ten different sections, allowing the angler to fish at a great range of lengths.

The tip sections of almost all poles contain elastic. This is tied directly to the line and acts as an absorber when the fish runs. Without it, either the line or pole would snap. Elastic comes in different strengths, depending on the size of fish (see page 19).

## roach poles

Roach poles are suited to traditional small-fish angling on rivers, canals and lakes for species such as roach, bream and skimmers. They are light, fine pieces of tackle and have flexible tips that are suited to elastics up to grade 8.

They, and most other poles on the market, can be taken apart and broken down from their full length so that they are fished at different lengths. This gives anglers the flexibility to add a section or two if they need to fish further out or take off a couple to come closer in.

## carp poles

As more waters across Europe began stocking carp, a different style of pole was needed to catch them. Lightweight roach poles were simply not up to the task of dealing with a hard-fighting species

**carp pole**

**roach pole**

**margin pole**

**whip**

that grows as large as carp do. Carp poles are strong, yet, thanks to their carbonfibre construction, incredibly light and capable of landing fish weighing well over 4.54kg (10lb). The extra strength comes from the thickness of the pole's walls, which are reinforced to guard against breakages. They also have a stiffer tip, which allows the use of stronger elastic, between grades 12 and 16.

## margin poles

This type of pole is for use when carp weighing into double figures are the target, and its name gives away its purpose: for catching in the margins, or nearside, of a lake.

Margin poles are shorter than carp poles – about 7m (23ft) on average – but they are strong and difficult to break. Often made of a blend of carbonfibre and glassfibre, they are ideal for taming big fish close to snags.

The downside of the pole's strength is the weight, but most margin fishing is done at a range of about 4m (13ft), so this isn't usually a problem.

## whips

The preserve of the small-fish angler, whips are unbeatable for catching nets of little fish, and they can, therefore, be regarded as specialized tools.

They are about 5m (16ft 5in) long and are like a pole but without the elastic tip. Instead, whips have extremely flexible solid carbonfibre flicktips, which act as shock absorbers. They are not much use for big fish, but are perfect for building up a good weight of small fish.

## how does a pole work?

If there was no elastic to act as a shock absorber, most fish would be lost when hooked on a pole. Elastic is available in different grades and should be matched against the species the angler wants to catch. Here's a quick guide to choosing the right elastic.

| grade of elastic | target |
|---|---|
| 1–6 | Ideal for catching small roach and perch with a roach pole |
| 7–10 | Perfect for catching bream and tench with a carp pole |
| 11–15 | Ideal for catching carp up to 3.63kg (8lb) with a strong carp pole |
| 15–20 | Designed specifically for carp up to 6.8kg (15lb) and best used with a margin pole |

# tackle sundries

## jargon buster

**Barbed hook** A hook with a small, sharp barb at the point.

**Barbless hook** A hook without any barb at the point.

**Shot** Balls of non-toxic metal with a small split, designed to be closed around the line to provide casting weight.

**Monafilament** The thin, flexible material used to make fishing lines.

**Breaking strain (BS)** The maximum weight a line can sustain without causing it to break.

**Hooklength** A short length of line that is attached to the main line at one end and the hook at the other.

After you've bought your rod, reel or pole, you'll need a number of other items before you can start fishing. Some are essential and these are described first, while the others will make fishing a more enjoyable experience for you.

## hooks

You can't catch a fish without a hook. Hooks come in different sizes, strengths and weights, depending on the type of metal they're made from. They come in even sizes and range from 2, the biggest, to 26, the smallest. Beginners, who simply want to catch their first fish, should opt for size 16 or 18 with double or single maggot.

Hooks can be barbed or barbless. Unhooking is easier with barbless hooks, and many fisheries insist on their use. Some hooks have an eye for attaching the line. Others have a spade end, to which line has to be whipped, and smaller sizes of these are used by some anglers (see page 41).

Anglers fishing for predatory species such as pike and zander use treble hooks, which are three hooks welded back to back to create a single hook. They are used to mount both live and deadbaits.

**the anatomy of a hook**

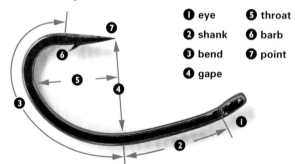

❶ eye  ❺ throat
❷ shank  ❻ barb
❸ bend  ❼ point
❹ gape

shot

line

swimfeeders

## shot

Lead shot was outlawed over 20 years ago in Britain because of its effect on wildfowl, especially swans, and a lead substitute was introduced, but in the rest of Europe you can still use it.

Shot is used to provide casting weight when floatfishing, and it ranges in size from the largest, SSG, which weighs 1.89g, to the tiny No.12, at 0.012g, for delicate pole rigs. It consists of a ball of metal with a split in it. Line is placed in the split and the two halves are pressed together to lock the line in place. When it's locked correctly, you can slide the shot up and down without damaging the line.

## line

Fishing line is made from monofilament (mono), a fine and flexible material that is available in a range of strengths and thicknesses. Monofilament is rated by the kilogram or pound, ranging from fine 0.45kg (1lb) line to thick and heavy 9kg (20lb) strength. For the beginner, 1.36kg (3lb) line will be sufficient for most circumstances.

Line is made to break at a certain weight – 1.8kg (4lb), for example – and will snap when a weight of that amount is pulled directly down on it. However, when it is used in conjunction with a flexible rod, and because fish are weightless in water, line will land a fish that weighs much more than its own strength. Line of 1.36kg (3lb) will easily land fish of three times that weight when it is used in the right way.

As well as main line, you can buy hooklength material, which is favoured by match anglers. It is a short length of monofilament, which is of lesser breaking strain (BS), with a hook attached. This helps you get more bites from shy-biting fish.

leger weights

floats

## swimfeeders

Swimfeeders or feeders are used to introduce bait into tight patches around your hookbait, and there are three basic models: open-end, block-end and method (see pages 34–5). They consist of a strip of perforated plastic, made into a tube and held in place with a strip of lead for casting weight.

## leger weights

Also known as bombs, leger weights are pear-shaped leads used as casting weights (see pages 34–5). They are designed to keep a bait motionless on the bottom and are most often used by anglers who want to get a bait further out than a float will cast or who are fishing for long periods of time.

A leger of about 7g (0.25oz) is fine for small fish, but for distance work, when you are after big carp and pike, you may need a bomb of up to 85g (3oz).

## floats

A float is a piece of material, usually peacock quill, plastic or balsa wood, designed for use as a bite indicator, and it is a far more visual way of fishing than using a swimfeeder or leger weight. There are four basic designs: straight crystal waggler, insert waggler, bodied waggler and stick float (see pages 32–3).

**tackle box**

**keepnet**

In addition to line, shot and hooks, there are several other items that the beginner may want to take fishing for the first time.

## tackle box

Tackle boxes are essential for keeping items like floats, hooks and shot safe and secure. These items would otherwise be left loose in the bottom of a bag or holdall where they could be easily broken or lost.

Generally made of toughened plastic, tackle boxes are available in a range of sizes, but almost all of them have separate compartments for each item of tackle. Choose one to suit the type of fishing you wish to do.

## landing net

When you catch that fish of a lifetime, the chances are that you won't be able to get it out of the water without a landing net, so you should regard this as an essential item.

The nets are made of a soft, knotless material so that they cannot harm the fish, but they vary in size depending on the sort of quarry you are after. You can get massive nets for big carp and tiny versions for roach on canals. Deep, triangular nets are best for big fish, while match and pleasure anglers prefer spoon-shaped, shallow nets. To see how best to use a landing net, study the guide on pages 100–101.

## keepnet

Nowadays keepnets are mainly used by match anglers, who retain what they catch so that everything can be weighed at the end of the contest. They are, however, also popular with people who fish for pleasure, especially anyone who likes to see what they've caught at the end of the day.

Modern keepnets are about 3.6m (12ft) long and are made from the same sort of material as landing nets.

## bait box

Purpose-made bait boxes are essential and should be on every beginner's 'must have' list before they start fishing. Baits such as maggots, casters, worms and sweetcorn can be safely kept in plastic bait boxes that have secure, tight-fitting lids. These lids have tiny holes in them that allow living bait to breathe but not escape.

Bait boxes are not expensive and are available in various sizes, holding from 0.5 to 2.8 litres (1–5 pints).

**bait box**

**seat box**

## plummet

A plummet is a small, inexpensive item, but one that is vital if you want to be successful. It is a piece of lead, which is attached to the hook for floatfishing so that you can determine the depth of the water in a particular swim. When you know this, you can decide what depth to present the bait, which is essential for targeting different species of fish.

## catapult

Feeding by hand is fine if you are only fishing a few metres (yards) out, but if you need to get bait out any further and feed the swim accurately, a catapult is essential. A small, match-style catapult is perfect for loose baits, like maggots, hempseed and pellet, but if you are using bigger items, such as boilies or even balls of groundbait, you will need a more powerful model.

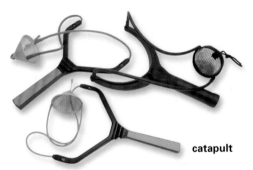

**catapult**

## discorger

Another inexpensive item that no beginner should leave home without is a discorger. This is a small piece of plastic that enables the angler safely to unhook a fish that is not hooked in the lip but deeper in the mouth (see page 102).

## seat box

Although a deckchair might suffice for the occasional angler, if you're serious about fishing, a seat box will prove a good investment. You will be able to store items such as reels, catapults and rod rest heads in a basic plastic box, but more expensive models have drawers and trays for hooks, line, floats, pole rigs and swimfeeders, as well as a deep section at the base for larger items like reels.

## luggage

Fishing luggage is not essential, but it will make transferring your tackle to the bankside and back much easier. Modern luggage is light, strong, waterproof and capable of carrying everything you will need for a day's fishing. Rod holdalls will not only carry your rods but will also hold banksticks, umbrellas and poles. A carryall will take tackle and bait boxes, together with nets and reels.

chair

rod pod

Once you've bought the must-have items for your first fishing trip, there are several other pieces of tackle that you might want to think about owning that will make fishing more enjoyable.

## chair

Some anglers find that a seat box is just too unwieldy and not comfortable enough for long fishing trips, and they find a custom-made, collapsible chair preferable. Most of these chairs are made from strong, lightweight metals and can be easily carried around.

## rod rest

You don't need a rod rest, but you'll hit more bites if you do have one. A rod rest is usually a V-shaped piece of plastic that screws into a separate metal stick, known as a bankstick, which can be set in the ground directly in front of you. You place your rod in the rest, keeping the butt on your knee.

## rod pod

When you gain experience and want to fish for larger quarry over longer periods, you might want to think about getting a pod. These are similar to rod rests, but overcome the problem of fishing from hard ground such as concrete and sun-baked mud. A pod is a collapsible metal or plastic structure that has rests for up to three rods. Some models have attachments for bite alarms and bobbins.

## bite alarm

These electronic alarms are popular with anglers fishing for carp, pike and other big fish. The alarms rests on a bankstick under the rod, usually between the handle and the first rod ring. The line running off the reel sits on top of a small, free-running wheel within the alarm, and when the line is pulled by a fish, the wheel spins and the alarm sounds.

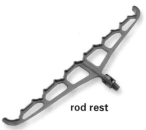

rod rest

bite alarm

bobbin

fishing umbrella

## bobbin

If you don't have a bite alarm or are legering for a short time, a bobbin will tell you if you have a bite. The bobbin clips on to the line between the reel and the first rod ring, and when a fish takes the bait it lifts up or drops down. Most bobbins are made of plastic, but washing-up bottle lids and silver foil work just as well.

## polaroid glasses

The glare from sun shining on the water can reduce visibility almost to nil, but Polaroid glasses help cut out the glare. They are particularly popular among specimen hunters and river anglers.

## forceps

When you go after bigger fish, such as pike and carp, a discorger may not be large enough, and only forceps will do. They are like surgical forceps: metal scissors with blunt ends. The ends are used to grip the hook shank and pull the hook out safely (see page 103). Forceps can also be clamped together to increase the hold on a particularly awkwardly placed hook.

## fishing umbrella

Fishing umbrellas are much larger and heavier than the umbrella you use to keep the rain off. They are usually green or black, and the better-quality ones are made from thick nylon, coated with a water repellent. They also have a strong, pointed pole, which can be stuck into the ground to stop the umbrella from blowing away. Most are at least 106cm (42in) across.

## scales

You will find it useful to carry a set of scales in your carryall – you never know when that big fish will bite and you want to weigh it. You can buy purpose-made scales from tackle shops, and they range in price from the inexpensive to expensive.

Only ever weigh a fish that is in a landing net or a special weighing sling, also available from tackle shops.

scales

# how do I ...?

# how do I
# ... spool a reel?

**Left** *The line on a spool should be right up to the lip to make casting easier.*

## jargon buster

**Backing** A material, such as electrician's tape, that is used to bulk out a spool in order to reduce the line capacity.

**Bedding in** The tendency of line that is wound tightly around backing on the spool of the reel to cut into the backing and make the reel appear less full.

Spooling your reel correctly will make casting much easier, cutting down on tangles and allowing you to reach feeding fish. The process is simple, but it's amazing how many beginners either over- or under-spool their reel. It's one of the most important things to get right before beginning to fish.

## line capacity

If you look at a reel, the line capacity will probably be marked on the side of the spool. This will tell you approximately how much line of a particular breaking strain (BS) or diameter will be needed to fill the reel spool neatly to the lip.

Match-type reels take about 100m (330ft) of light, 0.9–1.36kg (2–3lb) line, whereas reels designed for bigger fish, like carp and pike, will take a lot more line, sometimes up to 300m (985ft) of 4.54kg (10lb) mono.

Although this is only a guide, it is well worth trying to match up the line you've bought to the requirements of the spool. It will make the job of spooling much easier.

## backing

If you don't want to wind the full amount of line on to your reel, you can use a backing to bulk the spool out. There are several possible methods, but a popular one is to wrap electrician's tape around the spool until it is half-full. You can then wind line over the tape.

If you use tape you will probably notice over the first few sessions that the previously full spool becomes a little less full. This is because of what is known as bedding in. As the line is put under tension from casting and playing fish, it nestles into the tape a little. It is perfectly normal and you'll find that there's only a tiny difference between how full the spool was before and after.

## spooling a reel

Spooling your reel correctly will make casting much easier, cutting down on tangles and allowing you to reach feeding fish.

1 Remove the spool from your reel and fix the reel body to the butt section of your rod.

2 Pass the spool of line through the ring or rings on the butt section of the rod and tie the loose end to the detached reel spool, using a simple double knot. Make sure that this knot is situated at the bottom of the spool and not in the middle where it could hinder casting.

3 Once this is secure, open the bale arm on your reel and put the spool back on. Close the bale arm. You are now ready to start winding.

4 Ask a friend to hold the spool of line under tension to make sure that the line lies correctly on the reel as you wind.

# how do I
## ... learn to cast?

## jargon buster

**Bite** The moment the fish takes the bait.

**Strike** To react to a bite with an upwards or sideways movement of the rod or pole.

**Cast** The process of delivering your bait and rig to a specific spot.

Let's face it, you're not going to get far in fishing if you can't cast, and most beginners find it one of the most difficult things to master. As they say, practice makes perfect, so follow the step-by-step sequence provided and be prepared to practise hard.

## how to cast

Learning to cast can be a daunting prospect for beginners, but it is relatively straightforward. Study the four-step guide below and then practise on your local lake or river.

1 Make sure that you have the right amount of line between the rod tip and the float or swimfeeder. If the line is too short, the cast becomes jerky and will often not reach its intended destination. About 1m (3ft) of line is ample in most situations.

2 With the rod in a straight line directly behind you and the butt facing out into your chosen swim, hold the rod over your shoulder parallel to the ground. Hold the rod around the reel seat with one hand, trapping the line coming off the reel against the rod butt with your forefinger undo the bale arm with your other hand and grip the base of the rod handle.

3 Sweep the rod upwards from its position and follow it through until the tip is pointing out into the lake. As you do this, release the line trapped by your finger – do this roughly when the rod is coming over your shoulder. The compression built up by the rod through the cast will provide enough power and spring in the top sections of the rod to launch the float or swimfeeder to its destination.

4 The most important thing to master is judging when to let go of the trapped line. Too early and the line will spill off the spool at the wrong time; too late and the cast will be too short.

## how do I strike at a bite?

Once you've cast out, the next stage will be, ideally, to get a bite. Most beginners will start by floatfishing and a bite will be obvious – the float will disappear. When this happens, the angler must react quickly with a swift upwards movement of the rod. This needs to be firm and sharp to make sure that the hook is set – that is, that the fish has taken the bait and is caught on the hook. This movement is called striking.

If, on the other hand, you are legering with a quivertip, a simple, swift sideways movement will suffice. Many beginners strike too hard, losing their fish. Be calm but swift for success.

# how do I
# ... shot a float?

**Left** *An incorrectly shotted float is near useless, so take time to do it right.*

## jargon buster

**Dotting down** Shotting a float so that only a tiny portion of the tip is visible.

**Plummet** A conical piece of lead, often with a strip of cork in the base, used for finding the depth of a swim.

**On the drop** When fish such as roach and rudd rise to take sinking loose bait, they can be hooked on a slowly sinking hook. This practice is known as fishing on the drop.

Knowing how to get your float to sit correctly in the water – that is, how to shot it – is essential for floatfishing. Not only will it make casting easier and more accurate, but it will also get you more bites by allowing you to judge the correct time to strike.

Anglers often have too much float tip sticking out of the water, which means that a biting fish will have a job pulling it under. When the float carries the right amount of shot, you can do what is known as 'dotting down' so that only the top is visible. Then, when a fish bites, it takes little effort for it to submerge the tip and indicate a bite.

## positioning the shot

The amount of shot needed is marked on the side of most modern floats. A basic float rig should have the bulk of the shot 'locked' either side of the float, with smaller shot placed further down the line, depending on the species.

When you are after small fish, such as roach and rudd, which often feed away from the bottom and can be caught on the drop, spread out the shot so that the bait falls slowly through the water. For bottom-feeding species, such as bream and tench, place the shot together about two-thirds of the way down the line, so that the bait sinks straight to the bottom.

## finding the correct depth

Different species feed at different depths, so you must find the correct depth when floatfishing.

Estimate the depth and set up your rig, but attach a plummet to the hook before casting. The distance between the plummet and your float is your estimated depth. If the float sinks, you are too shallow; if it lies flat on the surface, you are too deep. Move the float along the line until you have found the correct depth. You can now determine where to present your bait.

**shotting a float**

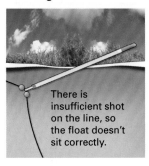

There is insufficient shot on the line, so the float doesn't sit correctly.

There is too much shot on the line and the float sinks.

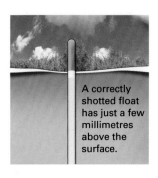

A correctly shotted float has just a few millimetres above the surface.

**Left** *These three pictures indicate what happens when a float is either under-shot or over-shot, or correctly shotted.*

## types of float

There are two main types of float: those attached through an eye, known as wagglers, and those attached by means of rubbers, known as stick floats. Stick floats are used for trotting on rivers, whereas wagglers are for stillwaters.

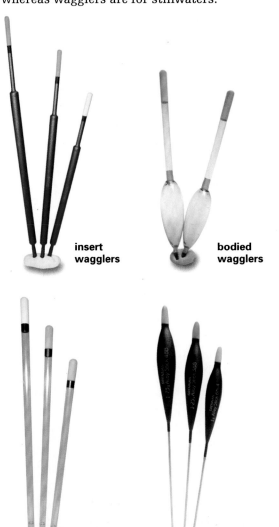

**insert wagglers**

**bodied wagglers**

**straight crystal wagglers**

**stick floats**

### insert wagglers

These are used on lakes, canals and slow-flowing rivers where there's less chance of the flow pulling the float under. Inserts have a plastic or balsa wood body with a thinner piece of plastic or wood at the tip. This creates a sensitive float, which is perfect for tackling fickle bream or for fishing on the drop for roach and rudd.

### bodied wagglers

These floats have a large body at the base with a balsa wood stem and tip. They are also made from clear plastic. These are used for casting long distances on big rivers and lakes. They often take more than four SSG shot (see page 21) and are used for bottom-feeding species, such as bream and carp.

### straight crystal wagglers

These relatively new floats now dominate the market. They are made of clear plastic with a brightly coloured tip, and they have the advantage of being invisible to fish. They are available in straight or insert styles.

### stick floats

Rather than being attached by the bottom end, stick floats are held on the line by three silicone float rubbers at the top, middle and bottom of the float. They are used on rivers for trotting through the swim and are usually made of balsa wood, which is suitable for most swims. However, floats with an alloy metal stem are useful for pacey, boily water in shallow swims, and sticks made from heavy lignum wood are good for casting at distance when the fish may be a little further out.

# how do I
# ... set up a leger?

## jargon buster

**Swivel** A movable small metal body with an eye at each end. It is used to attach main lines to hooklengths.

**Snaplink** A wire frame that can be unclipped to accommodate a feeder or leger weight. The clip is attached to a swivel.

When the wind is blowing strongly or the fish are feeding beyond the range of even the largest floats, most anglers will reach for the leger rod. This enables you to cast further and target bigger fish.

Legering involves the use of a leger weight (bomb) or a swimfeeder. The leger weight is designed for casting long distances (see page 21). Its use makes it possible to cast accurately beyond the range of the longest poles or biggest floats. However, it doesn't allow you to feed the fish, unless you use a catapult and groundbait, and this is a haphazard process for inexperienced anglers. This led to the creation of the swimfeeder, which can carry feed long distances.

There are two main ways of rigging a leger rig, and your choice will depend on your target species.

## types of lead and swimfeeder
### bomb
This is a pear-shaped lead with a swivel in the top, to which the line is attached. The aerodynamic shape means that a bomb can be cast a long distance, even in strong winds. Bombs are available in a range of weights.

### watch lead
Borrowed from seafishing, a watch lead is ideal in a flooded river when barbel are the target. It has grips that hold the bottom in fast-running water.

## link leger

A link leger is shot that is pinched to a piece of line coming off the main line. Typical shot ranges between BB and SSGs (see page 21).

## block-end feeder

This feeder is used to carry loose offerings or particles, such as maggots, casters, hempseed and pellets. It has a cap at each end to hold in the feed and an in-built lead strip to give casting weight.

## groundbait feeder

This plastic tube with holes is attached to a strip of lead, but it does not have caps at the ends. Before it is cast, the feeder is filled with groundbait and squeezed in at each end to make sure that the bait stays in place during casting.

## method feeder

This plastic frame has weight at one end or on the bottom, and it is designed to have a stiff groundbait moulded around it, creating a method ball. This is fished with a short hooklength, mainly for carp.

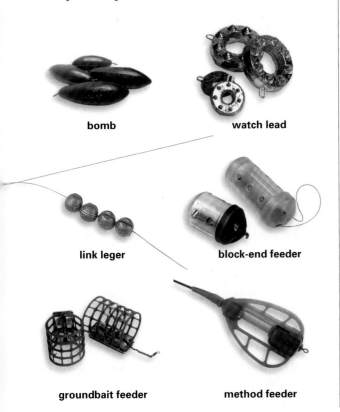

**bomb**

**watch lead**

**link leger**

**block-end feeder**

**groundbait feeder**

**method feeder**

## running rig

The sliding lead is preferred when chub or perch are the target. It is a swimfeeder or bomb that can run freely along the main line. This is a safe rig, because if the line breaks, the feeder will simply slide off the line rather than being towed around by the hooked fish.

1 Thread the feeder or bomb on to the main line and tie a swivel or leger stop to the end to act as a buffer for the lead.

2 Attach the hooklength and hook to the swivel.

## fixed rig

A fixed rig is used for docile species, such as bream and roach, which won't go charging off, running the risk of breaking the line. The most common fixed rig is a paternoster, which is especially popular among bream anglers.

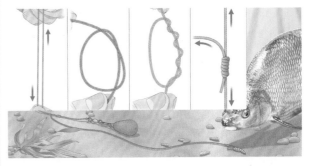

1 Tie a paternoster by forming a large loop in your main line – about 1m (3ft) is fine.

2 Decide how long you want your feederlink to be – about 25cm (10in) is normal. Cut the loop to produce this link and tie the feeder and snaplink to this tag.

3 Tie a small swivel to the end of the longer length of line and attach the hooklength and hook to this.

# how do I
# ... use a pole?

## jargon buster

**Unship a pole** To break down (shorten) a pole as the float is retrieved.

**Ship out a pole** To ease the pole out to its full length.

**Elastics** These range in thickness and are used to cushion against a running fish.

Although it looks like one of the oldest ways of fishing, the pole is a fairly recent introduction. Basically a length of carbonfibre with a line attached to one end, it is a lot more complicated than it appears at first sight.

The pole offers anglers supreme accuracy and presentation because the pole tip is directly above the float and there's little line for the wind to get hold of. Moreover, if you miss a bite, all you have to do is lift the rig out of the water and drop it back in to get fishing again.

### strike gently

Striking should be done in a careful, controlled manner with a simple lift of the pole. Use only enough power to set the hook. The elastic then takes over, stretching to subdue the fish. Once the fish tires, you can start unshipping and play it to the net.

### use the elastic

To land fish caught on a pole, you must use a length of internal elastic that runs through the top sections of the pole. The elastic comes in various strengths, and the gauge you use will depend on the size of fish you are after. No.1 is the lightest available, and it goes up to a whopping No.20 (see pages 18–19).

Because there is no reel, you have to push the pole gently out to the required distance. This is known as shipping out. Once you have hooked a fish, you need to bring the pole back, breaking down sections as you go, and this process is known as unshipping.

## pole floats

You can't use standard floats like wagglers and stick floats with a pole. Instead, you need some purpose-made floats, which tend to be much smaller and more delicate, largely because they are lowered into the water rather than being cast. These floats are attached by means of an eye on the side and a rubber, which is on a thin wire stem at the base.

## using a pole roller

To aid the shipping and unshipping processes, you will need a pole roller. Put this on the ground behind you when you are fishing and support the pole on it. It will also prevent the pole from being damaged by stones and mud.

## a simple rig

Pole fishing is an essentially simple method of fishing and as such is most effective when a basic float set-up is used. The diagram (right) shows how best to present a bait on a typical stillwater.

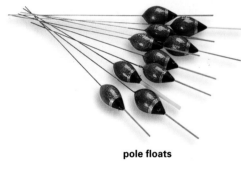

pole floats

pole roller

**a simple pole rig**

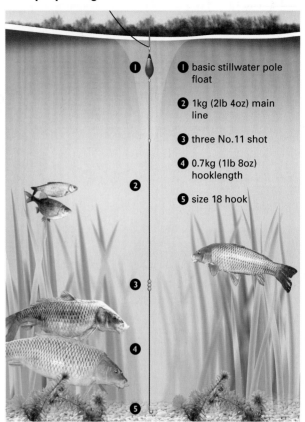

❶ basic stillwater pole float

❷ 1kg (2lb 4oz) main line

❸ three No.11 shot

❹ 0.7kg (1lb 8oz) hooklength

❺ size 18 hook

**Above** *The beauty of pole fishing is its simplicity. This rig is both easy to construct and effective.*

## playing a fish on the pole

1 Strike and let the elastic do its job. At this point keep the pole low and parallel to the water.

2 When the fish has stopped running, you can begin unshipping using the roller. Gently edge the pole back behind you, making sure that you keep a tight line to the fish at all times or it will come off.

3 When you have unshipped fully, break down the pole, still keeping that tight line. Most fish, even carp, should be easy to control from now on.

4 Slowly lift the top sections of the unshipped pole into an upright position, much as you would a rod. This will bring the fish up to the surface. If the fish runs, drop the pole back parallel with the water and wait for it to finish.

5 With the fish on the surface and beaten, slide it towards and into the waiting landing net.

# how do I
# ... mix groundbait?

**Left** *Groundbait is a brilliant and cheap way of encouraging fish into a swim.*

There's nothing better for getting a shoal of fish feeding in your swim than groundbait, especially for species like bream, carp and tench. In its most basic form it is relatively cheap to buy, but it also comes in a variety of different types. Suit the groundbait to the fish you wish to target.

Groundbait is material such as biscuits, breadcrumbs (made from bread dried in the oven and crushed), seeds and nuts that, when mixed with water, can be moulded into balls for throwing into your swim by hand or in a swimfeeder.

Mixing groundbait correctly requires care. If you add too much water, you will be left with an unusable mess that the fish won't like the look of.

Your aim is to make a fine mixture that makes a ball easily but breaks apart just as readily. It should not contain any large lumps or particles that might fill the fish up, which is especially important in winter when fish will want only small amounts of bait.

Follow the guide opposite to mix the perfect groundbait. Remember to take your time, as incorrectly mixed groundbait can be harmful.

# mixing groundbait

Never be in a rush to introduce groundbait. Instead, take your time and create the perfect mix.

1 Pour the dry ingredients into a large, round-sided bowl or bucket. If you are using more than one type, add them at the same time and mix them thoroughly by hand to make sure that they are evenly distributed.

2 Pour some water into a separate bowl and create a small well in the middle of the dry ingredients. Slowly pour water into the well, a little at a time, stirring as you pour so that the water is distributed evenly through the mixture.

3 When you think you have added enough water, use your hands to mix the groundbait to disperse any wet patches you haven't found. Leave the mixture to stand for about 30 minutes to make sure that all the water has been absorbed.

4 When it's ready, position a maggot riddle over another bucket and sieve the groundbait through the riddle into the empty bucket to remove any large pieces.

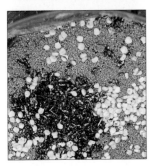

5 At this point you should add any maggots, casters, sweetcorn or hempseed that you want. The groundbait is now ready to use.

6 Either mould the groundbait into balls and throw into the swim or fill your swimfeeder with it.

## additives

Introducing liquid flavouring to your groundbait can make a huge difference to your catch – but how do you know what additive works best for which species? The table on the right suggests some of the best ones to use.

| flavour | target |
| --- | --- |
| Brasem | Bream |
| Strawberry | Roach, tench, carp |
| Scopex | Carp, tench |
| Vanilla | Roach, bream |
| Pineapple | Tench, roach |
| Tutti-frutti | Carp, bream, tench |
| Fishmeal | Carp, bream |

# how do I
# ... tie basic knots?

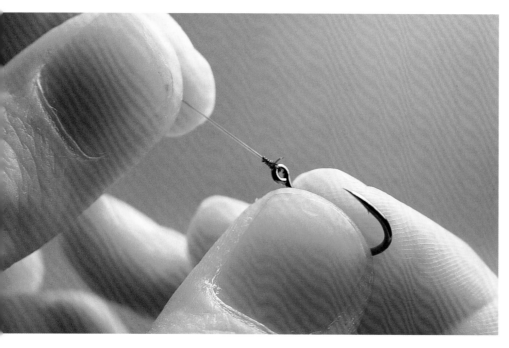

Where would we be without knots? They're probably the most important part of any tackle set-up, and a weak one could result in the loss of that prize fish. A mastery of the basic knots used in fishing is actually far more important than having the latest tackle or a super-secret bait. Five of the most often used knots are described here. They should cover almost every fishing situation you're likely to find yourself in.

## jargon buster

**Hooklength** A short length of line, lighter than the main line in use, attached to the main line at one end and to the hook at the other end.

## overhand loop knot

This is the knot most frequently used for creating loops to attach hooklengths.

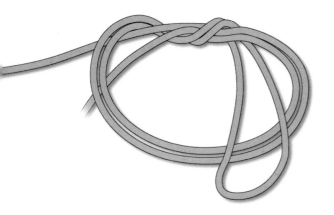

1 Double the line back on itself, creating a looped end. Hold this looped end and pass it back on itself and then over the doubled-up line below.

2 Pass the end twice through the loop you've just created. Pull the end tightly until the knot is secure and the line is tight. Trim the tag end.

## half-tucked blood knot

The half-tucked blood knot is used to attach swivels and eyed hooks to the line.

1 Pass the line through the swivel, hook or swimfeeder and bring it back on itself so that the line lies side by side. Twist the loose end around the main line between five and seven times, making sure that the main line is held tight.

2 Take the loose end and pass it through the little loop created between the swivel or hook and the first twist in the line.

3 Pull the loose end, moistening the knot with saliva to protect the line from damage. You may also need to push the knot together once it has been tightened to make it neat and tidy. Trim the tag end, leaving enough to compensate for any slippage during use.

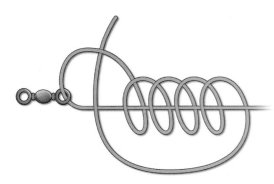

## sliding stop knot

This knot is often used when floatfishing in deep water or as a depth marker.

1 Take a 15cm (6in) length of line and place it next to your main line. Hold one end of the loose line and bring it back on itself to create a loop.

2 Pass this end through the loop several times, tying in the main line in the process.

3 Tighten the loose line, moistening it with saliva to protect the line. Once it is pulled tight, you should have a neat knot, which will slide up and down the main line. Trim the tag ends down to about 2.5cm (1in).

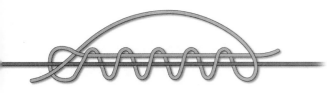

## water knot

This is an excellent knot for constructing a paternoster link when you are legering with a bomb or swimfeeder.

1 Place your hooklength material, which will normally be anything from 5 to 60cm (2–24in) in length, alongside your main line.

2 Make a loop and carefully thread the two lines through the loop four times.

3 Moisten the lines and pull tight. The tail from the main line should be used to tie on the swimfeeder or bomb. A hook should be attached to the hooklength line.

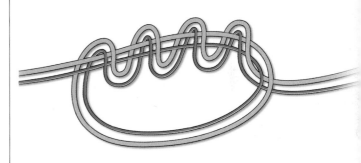

## spade-end hooks

Smaller sizes of spade-end hooks (see page 20) are often used by match anglers because they are believed to provide a more natural presentation. Here's how to tie a spade-end hook.

1 Make a loop in the line and lay it against the shank, leaving about 6cm (2½in) free at the end.

2 Wrap the end around the loop at least eight times and pass it through the loop. Remember to moisten it.

3 Hold the free end and apply steady pressure to the line so that the knot then slides tight.

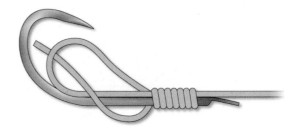

# which
# bait?

# which bait?
# bread

## key tips

### for using bread

**1** When you use flake always make sure that the hookpoint is showing, otherwise you'll miss bites.

**2** Breadpunch is a brilliant winter bait and should be used with mashed bread as feed.

**3** Try a floating piece of crust for carp in summer.

**4** If you haven't had a bite, strike off the bait so that you leave it in the swim as loose feed.

Bread is one of the best fishing baits because it is cheap and readily available, and is highly visible in the water. It can be legered or floatfished, and once it is immersed in water, it becomes soft and almost buoyant so that it wafts enticingly in the slightest current, which makes feeding fish less suspicious.

There are four main ways that bread can be used as bait: flake, crust, punch and mash. The first three are hookbaits; the last is used as feed.

## breadflake

Breadflake is taken from the soft insides of a fresh loaf. You can tear flakes of various sizes to suit the tackle you will be using and the species you want to catch. A large carp, for example, might be targeted with a piece of flake the size of a matchbox on a size 6 hook, while a fingernail-sized piece presented on a size 14 hook would be more appropriate for roach.

Fold the bread flake around the shank of the hook, leaving the end fluffy – but not so fluffy that it masks the point of the hook. Study the guide opposite to see how it's done correctly.

## bread crust

Breadcrust floats when it is put in water, making it a great surface bait for carp, chub and rudd in summer. At other times, crust will catch numerous species when it is fished at different distances off the bottom while legering or floatfishing. Small pieces can be combined with other hookbaits, such as maggots, to help fool wary fish.

## breadpunch

You can use a special-shaped punch to remove small pieces from a slice of bread, and this makes it possible to fish small pieces of bread on small hooks – sizes 14 to 18 – while you are targeting smaller fish, particularly roach.

Breadpunch is most commonly floatfished, either trotted under a stick float or with a pole.

*Above Mashed bread can be moulded into a ball and then thrown into the swim where it will explode!*

## mash

Thoroughly soaking and pulverizing stale bread creates a sinking feed called mash, which, when introduced, feeds a swim with lots of different-sized pieces of waterlogged bread and at the same time creates an attractive-looking, milky, cloudy effect in the water.

You can put stale bread through a food processor and the resulting white crumb can be squeezed into balls, which you can introduce to a swim by hand or press into a swimfeeder, from which it will explode after a few seconds of being in the water. This is excellent when used in conjunction with breadflake for chub in winter.

## flavoured bread

Bread can be easily flavoured, and there's no doubt that on some waters the fish prefer a scented bait.

The simplest way to flavour bread is to mount a piece of flake on your hook and add a couple of squirts of one of the commercially available bait sprays that you'll find in all tackle shops.

Alternatively, open the wrapper of a sliced loaf at both ends and use the same bait spray to cover both end slices. Refasten the wrapping, put the loaf in a bag and freeze it. When you take it out and leave it to thaw, the flavour will have permeated through the entire loaf.

Flavours range from cheese to pineapple, and you should experiment to find the ones that work best for you.

## breadflakes

A piece of breadflake is a bait that will encourage bites from a variety of species, and what's more, it's cheap and easy to use.

The key to ensuring that you use it properly is to mount it on the hook without masking the point, which will lead to missed bites. Follow this step-by-step guide.

1 Rip a piece of flake from the middle of a slice of bread and match it to the size of hook you are using. Place it behind the shank of the hook.

2 Gently fold the bread around the shank of the hook and squeeze it tightly. Leave the end of the bread in its 'fluffy' state.

3 Ensure that the point of the hook is clear to allow clear penetration when a fish picks up the bait.

4 The perfect breadflake should be squeezed around the shank to allow it to be cast but have an enticing fluffy end that doesn't mask the hook point.

# which bait?
# boilies

**Left** *Boilies have become very popular as a bait that captures specimen fish.*

## key tips

### for using boilies

**1** Shave the outside of the boilie to help release the flavour.

**2** Buy some buoyant boilies, known as pop-ups, which can be fished a few centimetres off the bottom to fool wary fish.

**3** Fruit flavours work in summer; fish-based flavours are better in winter months.

**4** Small boilies, known as mini-boilies, are preferable for tench and bream.

**5** Try dipping boilies in bait flavours to give you an edge.

Boilies have become popular in the last ten years or so. They are small, round balls of bait, available in many sizes, colours and flavours, which have been specially formulated with a mixture of powdered, high-protein ingredients. The resulting mix is bound with eggs, formed into balls and boiled (hence the name) to give a hardened outer layer. Boilies are often brightly coloured, with orange, red, white, yellow and cream being popular.

**Above** *Boilies come in all shapes, sizes and colours. They vary in flavours, too.*

There are two main types: shelf-life boilies and freezer baits. The shelf-life boilies contain preservatives to make them last longer. They are convenient to use and cheaper. Freezer baits are fresh and will go off if left at room temperature, but although they are more expensive, they are a better bait because they are fresher.

Some anglers prefer to acquire all the ingredients and roll their own, but the full range of colours, sizes and flavours is available from most tackle shops. For the beginner, though, it's easier to buy them ready-rolled in either a packet or a tub.

**Above** *Although boilies come flavoured, you can also buy additives that give them an extra 'kick'. These are best used on hookbaits only.*

## a big-fish bait

Boilies were developed for use in carp fishing and are often referred to as high-nutritional-value (HNV) or high-protein (HP) baits. They provide specimen fish with a long-term bait that is both good for them, making them grow bigger, and addictive, meaning that they will take them in preference to other baits, and they are a favourite with big carp, tench, bream and barbel.

The hard outer skin makes boilies more resistant to the attentions of smaller, 'nuisance' fish, so you can use them with confidence, knowing that only big fish will be able to eat them. Because they are hard, boilies are also easier to introduce into a swim by hand or catapult.

They can be as small as 1cm ($\frac{1}{2}$ in) or as large as 4cm ($1\frac{1}{2}$ in). As far as hookbait size is concerned, small boilies will catch fish of any size, but bigger ones will catch only those fish that are big enough to eat them. A variety of the normal boilie is what is known as a 'pop-up'. These are identical in size, shape and colour to a normal boilie, only these have a property in them that makes them buoyant. They can be presented at any distance off the bottom.

## using boilies on a hair-rig

Boiled baits are usually fished on a hair-rig presentation because they are usually too hard to mount straight on a hook. The advantage of the hair-rig is that because the bait is mounted on a hair of line off the front of the hook, the hook itself remains clear.

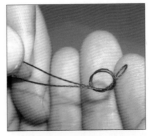

1 Cut your hooklength material 15cm (6in) longer than you want. Tie a small loop in the end of the line using an overhand knot (see page 40). This loop should be smaller than the bait you intend to use.

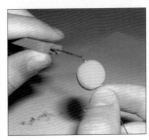

2 Use a baiting needle to pull the loop through the bait, then add a stop to keep it firmly in place.

3 Thread the line through the eye, going towards the point, and pull the hooklink through until the bait is lying about 2.5cm (1in) from the eye to the bottom of the bait.

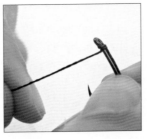

4 Take the tag end and whip the shank of the hook towards the bend. Always start whipping at the opposite side to where the gap in the hook lies.

5 Make eight whipping turns on the shank to hold the bait in place. Thread the tag end back through the eye, again towards the hook point, and pull tight.

# which bait?
# maggots and casters

Left *For sheer versatility, you can't beat maggots and casters.*

## key tips

### *for using maggots*

**1** Dead maggots can be a great loose feed because they don't wriggle away in the weed. Kill them by freezing them for at least three days.

**2** Maggots can be flavoured by adding commercially available flavours to a tub.

**3** In summer use floating maggots for rudd. Although the weight of the hook makes them sink, they do so slowly and enticingly.

Maggots and casters are two of the most popular fishing baits and are the best choice for beginners. They are cheap, readily available from tackle shops and nearly all species love them.

## maggots

Maggots are the larvae of the blowfly (bluebottle), and although you can breed your own, it's more convenient to buy them from a tackle shop. They are kept in a sawdust to stop them sweating and should be stored in a refrigerator to keep them fresh. Small fish, like roach, rudd, perch, dace and gudgeon, are routinely caught with a single maggot on size 20 or 18 hooks. Several maggots on a size 12 hook will often take bream, tench and carp.

### different colours

Visit almost any tackle shop and you'll see maggots in a whole range of colours. The natural colour, white, will certainly work on its day, but

Above *Casters are pupated maggots and tend to catch better fish.*

## guide to baits and hooksizes

The best way to present your bait correctly is to match it the hook size. Use this table as a guide.

| hook size | bait | hook size | bait |
|---|---|---|---|
| 24 | Single squatt | 12 | Breadflake; broken lobworm |
| 22 | Single pinkie | 10 | Small boilie |
| 20 | Single maggot; breadpunch; hempseed | 8 | Large lump of breadflake; bunch of brandlings; minnow livebait |
| 18 | Double maggot | 6 | Lobworm; boilie; cheesepaste |
| 16 | Redworm; triple maggot | 4 | Two lobworms |
| 14 | Small brandling; two grains of sweetcorn | 2 | Slug |

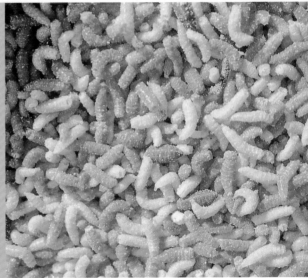

dyed grubs will give you an edge on different occasions and with different species. Red seems to be the most popular colour, and it's thought to work well because of its similarity to bloodworms (see page 51). Perch, roach, tench, carp and bream seem to prefer red. Bronze and yellow are also popular, and fluoro-coloured baits, which are bright pink, will also catch fish. Experiment by buying a mix of colours and find out which work best on your chosen fish.

### pinkies

Pinkies, the larvae of greenbottle flies, are smaller than normal maggots. They tend to be pink (hence their name) but are also dyed in a range of colours. They are best mounted on a size 22 hook. Match anglers rarely leave home without some pinkies because they are excellent for catching small fish. They are also ideal for use as loose feed when maggots are on the hook.

### squatts

The white maggots of houseflies, squatts are the smallest of the maggots and are usually offered as loose feed rather than as hookbait. In winter, when the fishing is hard, some match anglers use them on the hook, but they have to scale down to a size 24 to do so. They are far more effective when used in groundbait for bream, the species that seems to like them best.

### casters

The next stage of the maggot's metamorphosis into a blowfly is the caster (the pupated maggot). When maggots first turn into casters they are almost white, but as they age they become darker. Casters don't last long – they turn into flies after a few days – and they should be stored in the refrigerator. Casters should be used fresh before they become too dark, when they float and are useless.

They are a good alternative to maggots because, for some reason, they tend to catch bigger fish. You could be catching 60g (2oz) roach on maggots when a change to casters will produce a 115g (4oz) fish.

## hooking maggots and casters

Correct hooking is vital, and mis-hooking is a common mistake among newcomers. Take a maggot and gently squeeze it. You'll notice that a small tuft appears at the blunt end (where there seem to be two eyes) and this is where they should be nicked on the hook.

**Above** *Gently pinch the maggot at the 'eye' end.*

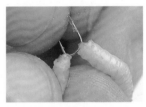

**Above** *Hook through the small tuft to leave the maggot free.*

# which bait?
# worms and other naturals

Left *Naturally brilliant, worms are one of the best baits – and they're free too!*

## key tips

### for using worms

**1** Nip off the end of a worm when you've mounted it on the hook. The fish will home in on the raw end.

**2** Worms can wriggle off the hook. A small piece of elastic band placed on the bend after the hook's gone in will keep it in place.

**3** Tread carefully when collecting lobworms because they are sensitive to vibration.

There is a multitude of natural baits to try that are not only readily available but also, more importantly, free. Worms, in all their guises, are best, but slugs, berries and shellfish also work well.

## worms

Worms are the finest of all naturals. They are plentiful and form part of the fish's natural diet because they are often washed into rivers and lakes by heavy rainfall, so all species are used to seeing them. There are several types of worm, however, and it's important to know the differences between them because each works for different species.

## lobworms

There isn't a fish that swims that won't eat a lobworm (earthworm). They are the largest of the three common worms, and you'll find them in your back garden where they can be dug up or collected from a wet lawn after dark. They are best stored in damp newspaper, which toughens up their skin and helps when it comes to casting.

## brandlings

These are much smaller than lobworms – up to about 10cm (4in) long – and they are red, with a soft skin that's marked with a series of yellow rings. Brandlings are found in manure and compost heaps, where they congregate in vast numbers, making collecting them easy. Keep them in the same sort of material they were found in.

redworm   brandling     lobworm

*Above Cut worms into tiny pieces to create a brilliant and effective fish attractor.*

*Above Try cutting a small piece of worm to trap the main bait on the hook.*

### redworms

Redworms are the smallest of the main types of worm and are particularly lively. They are similar to brandlings but are slightly darker red and have no yellow hoops. They live in compost heaps and should be stored like brandlings.

### bloodworms

Despite their name, bloodworms are the larvae of non-biting midges that live underwater. They are brilliant red in colour and tiny in size – barely a couple of millimetres long. They are a favourite with match anglers because they will catch fish when all else fails, and indeed, bloodworms are banned on many waters because they are so effective. They are expensive to buy.

## chopped worm

There is no better accompaniment to any worm hookbait than chopped worm. It is exactly as it sounds – tiny pieces of worm cut up with scissors to make a glorious but messy 'mush' that fish find irresistible.

Chopped worm can be introduced as loose feed or mixed in groundbait, but whichever way you use it, you'll find that it's a hugely effective way of enticing fish into the swim without filling them up.

## hooking worms

The keys to hooking a worm are to mount it so that it survives several casts and continues to wriggle and to make sure that the strike ends with a hooked fish.

Many beginners make the mistake of puncturing the worm too many times with the hook. This will kill it. Ideally, a lobworm should be threaded up the shank so that it is secure but still lively. Try using another small piece of worm on the hook to hold the main bait in place.

## other naturals

| bait | target | advantages and uses |
| --- | --- | --- |
| Slugs | Chub (especially free-lining) | • Collect from the garden and store in your bait box. <br> • Mount on a size 2 or 4 hook. |
| Elderberry | Roach | • Pick from early autumn to early winter. <br> • Loose feed three or four berries and fish one on a size 16 hook under a float. |
| Cockles | Tench (in summer) | • Use fresh from the fishmonger and leger or floatfish on a size 12 hook. |
| Prawns | Carp, tench and bream | • Dip them in a seafood sauce to make them even more attractive. |

# which bait?
# seeds, pulses and nuts

## key tips

### for using seeds, nuts and pulses

**1** Try flavouring your hempseed when you are boiling it, which can give you the edge in difficult conditions.

**2** Use coloured sweetcorn. Orange and red are great for tench and bream.

**3** Try including crushed tiger nuts or peanuts in your groundbait.

**4** Always make sure that the bait is properly and thoroughly cooked before fishing.

Seeds, pulses and nuts are used extensively in nearly all forms of angling, both as loose feed and hookbait. The most popular items are hempseed, sweetcorn, maize and tiger nuts, and although they are all available in tackle shops, you will save a lot of money if you buy in bulk from seed merchants or pet shops.

These baits are also known as particles because they are usually small and are mostly used in quantity in summer to entice non-predatory fish into a swim. They are particularly effective because they keep fish in an area without filling them up.

## preparing your bait

All of these baits, especially the bigger pulses and all nuts, must be prepared correctly or they may harm fish. Sweetcorn is the exception.

Almost all unprepared seeds, pulses and nuts must be soaked in water for 24 hours before being boiled for 30 minutes to make sure that they have absorbed as much moisture as possible. If it is not prepared, the bait will swell when it is in the fish's stomach, and this can be fatal.

Most tackle shops sell cooked hempseed and nuts, and inexperienced anglers should use these before they think about preparing their own. Remember always to ask for advice at the tackle shop if you're unsure.

**Above** *A brilliant summer bait, sweetcorn comes in a variety of attractive colours.*

| bait | uses | advantages | |
|------|------|------------|---|
| hempseed | • One of the best attractors for coarse fish, almost all of which are drawn to its smell and high oil content.<br><br>• Use as a hookbait for roach in cold weather when it should be floatfished on light tackle and a size 20 hook. | • A big bed of hempseed will encourage carp, roach, barbel, tench, chub and bream into a swim all year round.<br><br>• Because the seeds are so small, fish can spend a long time in a swim finding every piece. |  |
| sweetcorn | • During the summer this will catch almost all species.<br><br>• Use on the hook or as loose feed for tench, bream, chub and carp. | • It is inexpensive and readily available, both canned and frozen.<br><br>• It takes on colours and flavours well, which is useful when fish become wary of its plain, yellow form. |  |
| maize | • Used by specialist anglers who wish to target big carp. | • This is essentially giant sweetcorn, which is more attractive when it starts to ferment.<br><br>• Maize stays attached to the hook or hair-rig well. |  |
| tares | • Originally used for feeding pigeons, they make excellent hookbaits for catching roach. | • Best fished on the hook when hempseed is loose fed, but must be boiled before use.<br><br>• Tares tend to attract bigger fish than hempseed. | |
| tiger nuts | • This is an excellent hookbait for carp.<br><br>• Crushed or liquidized nuts make a superb addition to groundbait. | • Their attraction lies in their strong smell, which gets even stronger as they ferment, and in their high oil content. |  |
| peanuts | • They are a superb carp bait.<br><br>• Crushed into small pieces, they are worth adding to groundbait. | • Never use salted peanuts of the kind readily available in supermarkets.<br><br>• Boil them before use. | |

# which bait?
# pellets, paste and luncheon meat

**Left** *Pellets are brilliant as hookbait or loose feed.*

## key tips

### for using pellets, paste and luncheon meat

**1** Loose feed 3mm (⅛in) pellets and fish with a bigger one, say 5mm (¼in), on the hook. This is a great way of catching small carp.

**2** Add small pellets to groundbait. Bream especially enjoy pellets, which will keep them in the swim for long periods of time.

**3** A large lump of cheesepaste is a brilliant bait for chub in winter. The flavour will draw fish into a swim from a long way away.

As well as the traditional hookbaits, three other baits – pellets, paste and luncheon meat – will catch a range of species in the right conditions. Each require a slightly different approach and each will catch a multitude of species.

## pellets

Mainly made of fishmeal and fish oils, pellets range in size from 3mm (⅛in) to 2.5cm (1in) across. They have become increasingly available in recent years as aquaculture (the farming of fish for food) has become more common and pellets are used to feed farmed fish and make them grow bigger. When they are immersed in water, pellets start to go soft and give off flavours and oils. This, combined with the fact that many of the fish stocked into fisheries were fed pellets early in their lives, makes pellets attractive to fish.

## using pellets

Pellets contain high levels of protein and oil, and anglers should use them cautiously during the colder months of the year. Because they are cold-blooded, fish digest food fairly quickly in warm weather, so the amount offered as loose feed isn't important. In winter, however, a fish's metabolism can slow down so much that if they are given too much pellet, they might fill up and not need to feed for days afterwards, which can affect your chances of catching. Therefore, introduce pellets sparingly in winter.

## cheesepaste

One of the best winter baits for chub is cheesepaste. It's cheap and easy to make too.

1 Crumble up some cheese – Stilton is great because it has a very strong smell – and wrap it in a piece of shortcrust pastry.

2 Wrap the cheese in the pastry and mould it into a ball. Simply tear a piece off and put it on the hook. Experiment with different cheeses until you find the one the fish you are targeting like.

The most common pellets are for halibut, carp and trout. They are usually hard and should be attached to the hook by means of a hair-rig (see page 47). Soft pellets are available, as are pellets that are bought dry and then soaked in water, and both can be put directly on the hook.

## pastes

Paste is a superb year-round fish-catcher, and all non-predatory fish eat it. You can make your own from almost any ingredients that are attractive to fish, allowing you to come up with a unique, tailor-made bait that no other angler is using and that fish have little reason to suspect.

### types of paste

Ready-mixed paste can be bought in tackle shops, and the ingredients are similar to those used in boilies (see pages 46–47). Paste can be red, brown or yellow and can be sweet or flavoured of fish.

Bread paste is perhaps the most popular of the homemade mixtures, and it can be made by kneading slightly stale bread while slowly adding water until you get the right consistency. The other popular mix is cheesepaste, made by combining stale bread and grated cheese or a strong cheese, such as Danish Blue, and fresh shortcrust pastry (available ready-made from supermarkets). Margarine or a little cooking oil can be added to make the paste softer if necessary.

### using paste

Pieces of paste are broken off and moulded around the hook. Soft paste is better than a hard mix because it will exude more enticing flavours, so the most important aspect is to keep it on the hook. The pole is, therefore, probably the best rig because it allows you to place the bait gently and accurately. The paste lasts for a few minutes before it must be replaced, but the bonus is that you are feeding the swim.

## luncheon meat

Pork-based luncheon meat is a superb bait for all coarse fish, especially barbel, chub, carp, tench and bream. However, all processed meats can be used successfully, including sausages, meatballs, frankfurters, garlic sausage and corned beef.

### using luncheon meat

The meat can be mounted straight on to the hook or used on a hair-rig. For big fish, like barbel and carp, cut a piece the size of a matchbox and put it directly on a size 4 or 6 hook. If you are after tench and bream, use a smaller piece and a size 14 hook.

Flavour bait-sized pieces by frying them with chilli powder or put the pieces in an airtight container with your chosen flavour and freeze them in the freezer.

**Above** *Luncheon meat can be cut into a variety of sizes, from large cubes to a fine mince.*

# which bait?
# deadbaits and lures

## key tips

*for using dead-baits and lures*

**1** When you are fishing for predators, always use a wire trace – a short piece of thin metal to which you attach a lure. It will prevent predators, which have sharp teeth, from cutting through the main line.

**2** Try puncturing deadbaits to release their attractive juices.

**3** Spend five minutes retrieving a lure through a swim, then move on. That way you cover a lot of ground.

**4** Match the size of the lure to the fish you want to catch. Smaller ones work for perch; bigger ones for pike.

**5** Carry a pair of forceps when you're targeting predators. You'll need them to remove big hooks from sharply toothed mouths.

Predatory fish – pike, zander, catfish, chub, eels and perch – can be caught with deadbaits or lures. While deadbaiting tends to be static, lure fishing requires the angler to be mobile.

## deadbaits

There is such a wide choice of deadbaits available that fewer and fewer anglers are using live fish, and they are probably best avoided altogether by newcomers. Deadbaits are, in any case, much easier to store.

All tackle shops sell a variety of deadbaits, but to get the freshest available, simply catch them from the venue you intend to fish and humanely kill them before putting them on the hook. Seafish, especially mackerel, smelt and lamprey, also make excellent deadbaits.

## coarse fish deadbaits

| bait | target | advantages and uses |
|---|---|---|
| minnows | Big perch | • Catch them in big landing nets from the margins of lakes and rivers, and use immediately. <br> • Lip-hook on a size 8 hook and present under a float. |
| roach | All predatory species; zander prefer smaller roach; pike prefer bigger ones | • Fish weighing 60–115g (2–4oz) are the staple diet of all predatory species. <br> • Anything bigger than a small minnow should be mounted on a treble hook. |
| skimmer bream | Pike, zander and catfish | • Use fish weighing 60–115g (2–4oz) and mount them on a treble hook. |
| mackerel | Pike | • Use whole or, preferably, in sections because they are an oily fish, leaking enticing juices underwater. <br> • Good for long-range casting because they are firm and stay on the hook. |
| smelt | Pike and zander | • Readily available from fishmongers. <br> • Underwater they give off a strong smell and their pale colour makes them highly visible. |
| lamprey | Pike, chub and catfish | • Cut into sections so that the juices are exuded underwater. |

# lures

Lure fishing is becoming increasingly popular across Europe because it requires the minimum amount of tackle and allows anglers to be mobile. All you do is cast a spinner, spoon, plug or jig and then retrieve it. Lures are designed to mimic prey or to encourage the fish to attack their vivid colours or noise, and there are four basic patterns.

## spinners

These are so called because they have a blade that rotates around a central post. Fish are attracted by the bright, shiny colours and also by the underwater vibrations, which alert the attention of predators. Perch are the main species to fall for this type of lure.

## spoons

Spoons are bigger than spinners and do not have a separate spinning body, being made of solid metal. They are usually coloured silver or gold and can look dramatic when they catch the sun. Pike and perch fall to spoons. Remember to use a wire trace – if a predator cuts through the line and is left with a lure in its mouth, it could be fatal.

## plugs

These are perhaps the most attractive of lures and those that mostly resemble fish. They are effective because they can be retrieved through water at a specific depth, which is determined by the plastic 'lip' they have at their 'nose'.

## jigs

These relatively new pieces of kit get their name from the way they move across the bottom when they are retrieved. They are made of rubber and have just a single hook, which is weighted to reach the bottom. The large rubber 'dress' they boast provides brilliant visual stimulation, and jigs are excellent for catching perch and pike.

**Above** *A selection of plugs. Notice their resemblance to fish and their vivid colours.*

know your
fish

# know your fish
# barbel (*barbus barbus*)

**coloration**
The back is brown, the flanks are a bronze-copper colour and the belly is white.

**dorsal fin**
Pointed in shape, it is set in the middle of the back and is dark brown.

**head**
The head is large with a broad snout. The eyes are yellow.

**tail fin**
This is large and forked and it is where the fish gets its power from.

**scales**
The scales of the barbel are small and situated on the outer skin, giving it a rough texture.

**mouth**
The mouth is large, and there is a pair of short barbules at the front of the mouth and a pair of longer ones at the sides. These help during feeding.

factfile

A barbel can live for more than 15 years.

Barbel love to 'flank', a term used to describe the way in which they twist their bodies while they are feeding.

They fight hard, using their body shape in the fast water, so stout tackle is needed.

Like pike, they are especially vulnerable in summer, so take care in returning them, allowing them to rest for ten minutes in the landing net if necessary.

Barbel are often confused with gudgeon, but they have four barbules, while gudgeon have just two.

The slim and streamlined barbel is perfectly designed for its river habitat. It has a distinctive long, lean shape, with a pointed head and underslung mouth. It also has four sensitive barbules – two small ones at the tip of the nose and two longer ones at the sides of the mouth – which are used to scour the gravel for grubs.

## how to recognize them

With their golden-bronze flanks, dark fins and white underbellies, barbel are widely regarded as one of Europe's most beautiful fish. They can grow to more than 9kg (20lb) in some European countries, but the average weight is nearer 2.27kg (5lb), and this can be regarded as a good fish. A specimen would be one of 4.54kg (10lb).

## where to find them

Barbel are found in most areas in west and central Europe, including southeast England and south Wales, east to Russia and the Black Sea. They are absent from Scotland and Ireland.

They are almost exclusively found living in rivers. They love fast-moving water, especially areas where the pace is so strong that silt cannot build up on the bottom, leaving clean gravel instead. This is the prime area to find barbel. The quality of the water is also important, and the species thrives in rivers with high oxygen levels. Weirs are another barbel hotspot because of the constant influx of fresh water.

Barbel spawn over clean gravel in spring, when vast numbers head to shallow areas where females create a small indentation in the gravel to lay their eggs, which are then fertilized by the males.

You can also catch barbel in stillwater, although these venues are few and far between. The fish have been artificially stocked, but although they are away from their natural environment, they appear to thrive.

## what do they feed on?

The natural diet includes insect larvae, snails and freshwater mussels. Barbel have also been known to eat small fish, although this is less common. Feeding times are dictated by the weather. In summer they will eat at dawn and dusk, with the first few hours of darkness being especially favoured. In winter, when the water is cold, they are less active, but if the river is in flood, they have been known to feed heavily.

## how to fish for them

Both float and leger tactics work for barbel, but in rivers where the flow is strong, legering will be more effective. On small rivers, where it's possible to feed by hand, a straightforward leger weight will be best, but on large swims, use a swimfeeder. Floatfishing on rivers will work well, but this is a skill best left to more experienced anglers.

Barbel love hempseed, and this is a great bait for getting them into the swim and holding them there. When they are occupied with these tiny grains, you can offer sweetcorn, luncheon meat or boilies.

In winter, if the water is coloured after flooding or a storm, use a smelly, oily bait like luncheon meat to attract their attention. Don't forget to try a flavouring on your bait – meaty flavours work best in cooler temperatures.

## six baits to try

1  Luncheon meat
2  Maggots
3  Worms
4  Boilies
5  Sweetcorn
6  Pellets

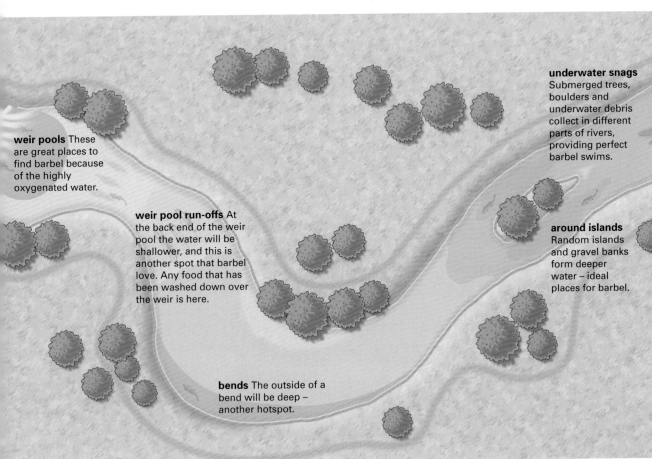

**weir pools** These are great places to find barbel because of the highly oxygenated water.

**weir pool run-offs** At the back end of the weir pool the water will be shallower, and this is another spot that barbel love. Any food that has been washed down over the weir is here.

**bends** The outside of a bend will be deep – another hotspot.

**underwater snags** Submerged trees, boulders and underwater debris collect in different parts of rivers, providing perfect barbel swims.

**around islands** Random islands and gravel banks form deeper water – ideal places for barbel.

# know your fish
# bream *(abramis brama)*

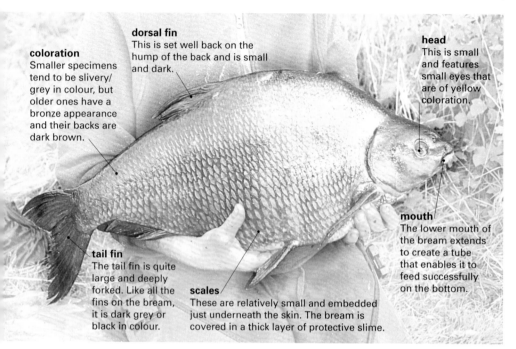

**coloration**
Smaller specimens tend to be silvery/grey in colour, but older ones have a bronze appearance and their backs are dark brown.

**dorsal fin**
This is set well back on the hump of the back and is small and dark.

**head**
This is small and features small eyes that are of yellow coloration.

**mouth**
The lower mouth of the bream extends to create a tube that enables it to feed successfully on the bottom.

**tail fin**
The tail fin is quite large and deeply forked. Like all the fins on the bream, it is dark grey or black in colour.

**scales**
These are relatively small and embedded just underneath the skin. The bream is covered in a thick layer of protective slime.

Common bream are one of the most widespread of all coarse fish as they live in a range of different water types, from lakes and gravel pits to drains, canals and rivers. They move in vast shoals, sometimes hundreds strong, lazily grazing over gravel and silt beds in search of food. They are a popular species with anglers, primarily because of their shoaling habit, which means that if one is caught, it is more than likely that several others will follow.

## how to recognize them

Bream have deep, narrow bodies, and it's this shape that has led anglers to call small fish skimmers, while bigger specimens – those over 1.36kg (3lb) – are known as slabs. Younger fish tend to have a silvery appearance, but older ones are a darker bronze-brown, and the common bream is also known as the bronze bream.

The average weight is about 1.36kg (3lb), and although they have been known to grow close to 9kg (20lb), anything approaching 3.63kg (8lb) can be considered a specimen to be proud of. As with all fish, don't set your sights too high.

## where to find them

Bream are widespread across Britain and Ireland, and can be found in France and most of continental Europe.

They love wide open spaces, especially in lakes and gravel pits, where they graze before moving on to another feeding area, and they are nomadic fish, covering vast distances in search of food. Unlike tench, which like to patrol the margins of a lake, bream are much more likely to be found further out, normally around underwater features, such as gravel bars, depressions or plateaux.

**gravel bars** As they travel around the pit in search of food, bream will cruise over the top of gravel bars.

**gullies and drop-offs** Look for bream in some of the deepest areas of the gravel pit, like gullies and drop-offs.

When they are in the feeding mood, which is usually at first light or dusk, they will show themselves by rolling on the surface or sending up patches of feeding bubbles. Tench also do this, but bream bubbles tend to be bigger.

They spawn in late spring or early summer, when the water is beginning to warm up. Males are easy to recognize at this time because they grow hard spawning tubercles around their heads and bodies. Each female lays as many as 400,000 eggs, but only a fraction of these survive.

## what do they feed on?

In their natural environment bream feed on a wide range of plant material, insect larvae and other bottom-dwelling insects, but they will eat almost anything put in front of them.

## how to fish for them

Bream respond particularly well to groundbait. If an angler introduces lots of breadcrumb by means of a swimfeeder, this will keep a shoal of bream in the area. Sweetcorn, bread, worms, maggots, casters and small boilies are all great baits for bream.

## silver bream

The silver bream *(Blicca bjoerkna)* is often confused with small, skimmer bream, but it is a completely different species. It is much smaller – rarely growing more than 0.45kg (1lb) in weight – and has a large eye and grey-tipped, reddish fins. As the name suggests, its flanks are silvery-white, but it has no slime on its body, unlike the common bream. Expect to find silver bream in weedy ponds and lakes, as well as canals and rivers, where they feed on snails, bloodworms and midge larvae. It is widely regarded as much rarer than the common bream.

## six baits to try

1  Lobworms
2  Sweetcorn
3  Bread
4  Groundbait
5  Pellets
6  Maggots

# know your fish
# wels catfish (*silurus glanis*)

This ugly species, also known as the European catfish and Danubian catfish, is one of the largest of the freshwater fish, with specimens as long as 3m (10ft) having been recorded. It is said by some to resemble a giant slug.

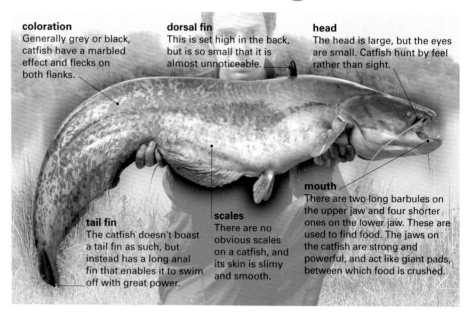

**coloration**
Generally grey or black, catfish have a marbled effect and flecks on both flanks.

**dorsal fin**
This is set high in the back, but is so small that it is almost unnoticeable.

**head**
The head is large, but the eyes are small. Catfish hunt by feel rather than sight.

**tail fin**
The catfish doesn't boast a tail fin as such, but instead has a long anal fin that enables it to swim off with great power.

**scales**
There are no obvious scales on a catfish, and its skin is slimy and smooth.

**mouth**
There are two long barbules on the upper jaw and four shorter ones on the lower jaw. These are used to find food. The jaws on the catfish are strong and powerful, and act like giant pads, between which food is crushed.

## how to recognize them

They have dark brown bodies, marbled with grey, and tiny dorsal fins. The huge mouths have a protruding lower jaw, which acts like a pad to crush its prey. Perhaps their most striking characteristic is the whiskers. There are two long barbules on the upper jaw and four shorter ones on the lower jaw, and the fish use these to feel for prey on the bottom.

They can grow large, and some have been recorded at more than 90kg (200lb). Even the average weight is big – fish less than 6.8kg (15lb) are rarely seen. There are lots taken each year weighing more than 45kg (100lb), and you should consider a catfish weighing 9kg (20lb) a good catch and one three times that weight a specimen.

## where to find them

Wels catfish are found in freshwater throughout eastern Germany and Poland as well as in France, Spain and some areas in southern England. They prefer gravel pits, ponds and big, wide, slow rivers, but are less likely to be found in canals and fast-moving rivers.

## what do they feed on?

Catfish eat other fish, both alive and dead, but they have also been known to take waterfowl, as well as small mammals and frogs. Although normally bottom-feeders, they will occasionally take prey near the surface of the water. They are regarded as scavengers and do most of their feeding after dark.

## how to fish for them

Legering is probably the most effective method, largely because catfish feed at night when a float is virtually useless. Catfish fight hard, so you will need strong rods, heavy lines and big hooks. The best baits are livebaits – roach and carp are favourites – although a bunch of lobworms or a fish-flavoured boilie will also work. It pays to experiment and find out what works well in your stretch of water.

## six baits to try

1  Roach livebait
2  Carp livebait
3  Fish-flavoured boilie
4  Bunch of lobworms
5  Luncheon meat
6  Pellets

# know your fish
# dace *(leuciscus leuciscus)*

This relative of the chub is one of the smallest species likely to be targeted by coarse anglers. It is usually found in shoals, like other cyprinids.

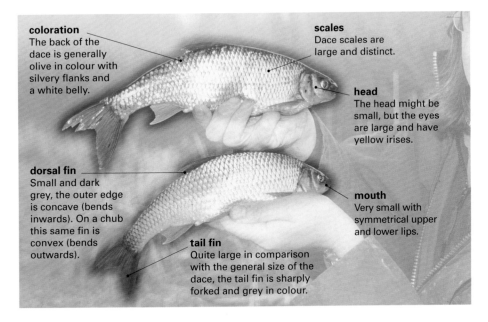

**coloration**
The back of the dace is generally olive in colour with silvery flanks and a white belly.

**scales**
Dace scales are large and distinct.

**head**
The head might be small, but the eyes are large and have yellow irises.

**dorsal fin**
Small and dark grey, the outer edge is concave (bends inwards). On a chub this same fin is convex (bends outwards).

**tail fin**
Quite large in comparison with the general size of the dace, the tail fin is sharply forked and grey in colour.

**mouth**
Very small with symmetrical upper and lower lips.

## how to recognize them
It is a small fish with a silver-coloured body, yellow eyes, a dark grey dorsal fin and yellow-orange anal and pelvic fins. They are sometimes confused with small chub, but they can be quickly distinguished by the fact that the dorsal fin of the dace is concave (bends inwards), whereas on a chub it is convex (bends inwards).

Dace rarely exceed 0.45kg (1lb) in weight, so individual targets are largely pointless. They move around in huge shoals, often numbering several thousand, and often anglers can spend hours catching fish weighing 30–60g (1–2oz), and a bag of 2.27kg (5lb) will probably include 100 individual fish.

## where to find them
One of the most common of Europe's fish, dace can be found in waters all across Europe, except for the extreme south and north of the continent. They prefer clean, fast-flowing rivers and streams with gravel beds. Although they occasionally turn up in stillwater, it is rare and almost certainly because they have escaped during flooding.

## what do they feed on?
Dace aren't fussy eaters, taking almost anything they can find. Shrimps, algae, floating insects and small snails will be on the menu, and they are as happy feeding off the surface as they are the bottom. As far as angler's baits go, opt for a small bait like maggot, caster or breadpunch.

## how to fish for them
Floatfishing is the best – and quickest – way of catching dace. Trotting, which is the art of allowing a float to run through a swim with the current, is perhaps the most effective, with maggot the preferred bait.

A small swinfeeder will also work well, but if you want to leger for dace, a small link leger with a light lure and a small hook is best.

## six baits to try
1 Maggots
2 Casters
3 Breadpunch
4 Hemp
5 Redworms
6 Elderberries

# know your fish
# common carp *(cyprinus carpio)*

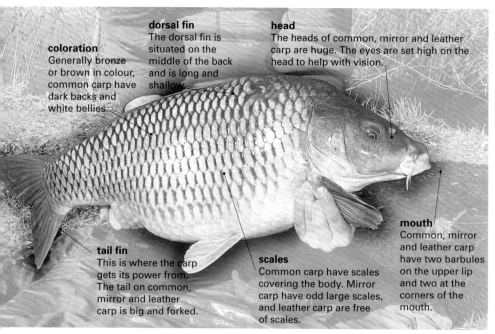

**coloration**
Generally bronze or brown in colour, common carp have dark backs and white bellies.

**dorsal fin**
The dorsal fin is situated on the middle of the back and is long and shallow.

**head**
The heads of common, mirror and leather carp are huge. The eyes are set high on the head to help with vision.

**tail fin**
This is where the carp gets its power from. The tail on common, mirror and leather carp is big and forked.

**scales**
Common carp have scales covering the body. Mirror carp have odd large scales, and leather carp are free of scales.

**mouth**
Common, mirror and leather carp have two barbules on the upper lip and two at the corners of the mouth.

## factfile

Carp can live for a long time, with fish known to be more than 40 years old.

Although mostly found on the bottom, carp love to cruise around under the surface in summer where they eat insects from the surface.

Carp love to feed after dark so night fishing is worth trying.

Use strong tackle because carp fight very hard!

Carp are among the fastest growing species and can reach 0.9kg (2lb) in a year.

The common or king carp is one of Europe's most popular coarse fish, largely because of its fighting abilities and its strength. River-dwelling carp tend to be more streamlined and athletic-looking than their stillwater-dwelling cousins, and a fish of 4.54kg (10lb) will give you a taste of how hard these fish can fight.

## how to recognize them

The common carp is covered in scales (apart from its head), but you are also likely to encounter mirror carp, which have a random scattering of scales, and leather carp, which are smooth and scaleless. All three have the same broad, deep shape and golden-brown colour. They also have four barbules around their mouths, which they use to taste food.

Carp grow to enormous sizes, with fish weighing more than 32kg (70lb) possible. You are more likely to catch a 13.6kg (30lb) fish now than ten years ago, although a fish in excess of 9kg (20lb) can be regarded as a specimen. Carp weighing more than 18kg (40lb) are caught during every season.

## where to find them

Carp are found in freshwater in southeastern Europe and have been introduced to waterways throughout Europe except for Iceland and northern Scotland. They are more likely to be found in ponds, lakes and gravel pits than in canals and fast-moving rivers. The common and mirror carp can be found throughout Europe, but the leather carp is much rarer.

They like to inhabit water that is rich in plant life, where they lazily root around the bottom in search of food.

When carp spawn depends on the temperature, but late spring and early summer are usual, because the water is beginning to warm up. The female lays over a million eggs among the weeds

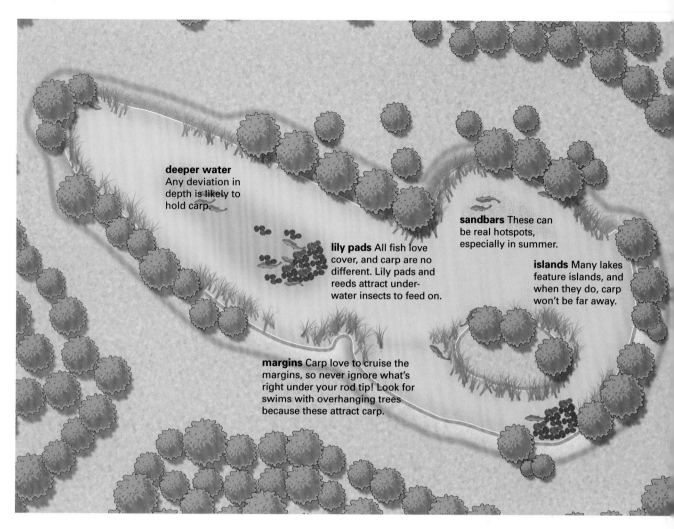

**deeper water** Any deviation in depth is likely to hold carp.

**lily pads** All fish love cover, and carp are no different. Lily pads and reeds attract underwater insects to feed on.

**sandbars** These can be real hotspots, especially in summer.

**islands** Many lakes feature islands, and when they do, carp won't be far away.

**margins** Carp love to cruise the margins, so never ignore what's right under your rod tip! Look for swims with overhanging trees because these attract carp.

in the shallows, and when they hatch, the surviving fry rapidly become fish.

## what do they feed on?

Carp will eat snails, bloodworms, mussels and anything else they can find in the mud on the bottom. They have large mouths and spend hours sucking and blowing out morsels of food – sending large bubbles to the surface in the process. Although they are mostly bottom-feeders, they will also come to the surface of the water and take small insects.

## how to fish for them

The hours of darkness are perhaps the best time to catch carp, although they also feed at dawn and dusk. You can floatfish or leger, depending on the water you are fishing. On small ponds a float is an excellent way of presenting a bait, but if you want

to try and catch a really big fish – 6.8kg (15lb) or more – legering is the better option.

Look for swims that have features such as overhanging trees, islands or underwater gravel bars because carp love to move around these areas.

Carp are at their most active during the warmer summer months, but they can be caught in winter when cold snaps are broken by a small rise in the temperature.

Choose something like a boilie, paste, lobworm or sweetcorn, but if the fish are feeding on the surface, floating crust is a great alternative. Remember to be patient when fishing for carp as the big ones don't show up very often!

## six baits to try

1 Boilies
2 Bread
3 Pellets
4 Sweetcorn
5 Dog biscuits
6 Lobworms

# know your fish
# chub *(leuciscus cephalus)*

**mouth**
Chub have extremely large mouths and thick lips. The jaws are also strong, allowing the species to eat large, hard food items.

**dorsal fin**
Set in the middle of the back, the dorsal fin is small and convex (bent outwards). This helps differentiate between small chub and dace.

**tail fin**
This is large and dark grey in colour. It is also forked.

**head**
The head is large and the chub has a blunt snout. The eyes have a yellow iris.

**scales**
These are large and prominent on a chub.

**coloration**
Chub have a dark grey back with silver-bronze flanks and a white belly.

factfile

In summer, when the water is clear, it's possible to stalk dace by creeping up on them and then watching them take the bait.

Chub are traditionally a river species but they can also be found in stillwaters.

They have huge mouths for their size, and a 0.9kg (2lb) fish can easily eat bait the size of a golfball.

In summer they love to bask in sunshine and will take flies off the surface.

Often when no other species will feed in the depths of winter, you can still catch chub.

Chub are known as shy and they love the cover of overhanging trees, weedbeds and undercut banks. In summer, however, they come out from their hiding places and bask under the sun. At times like this the angler must be stealthy, for the fish will dart for cover if they are disturbed.

## how to recognize them

Small chub can be mistaken for dace, although they have different dorsal fins (see page 65 and right). Bigger specimens can be easily recognized by their brassy coloured flanks, orange anal fins and huge, white lips, which can engulf big baits. Stillwater chub can tend to be less streamlined and more barrel-shaped than their river-dwelling cousins.

Although they are not one of Europe's biggest freshwater fish, chub can grow to 4.54kg (10lb) in weight. However, the average size in Britain is much more likely to be 0.9–1.36kg (2–3lb), and one weighing more than 2.27kg (5lb) is a specimen. However, giants of 3.63kg (8lb) are caught on the rivers of continental Europe every season.

### How to tell the difference between a chub and a dace

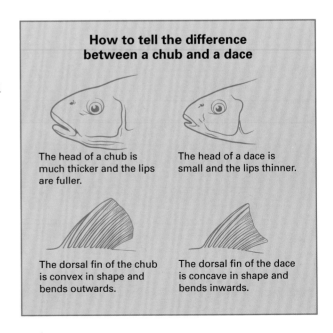

The head of a chub is much thicker and the lips are fuller.

The head of a dace is small and the lips thinner.

The dorsal fin of the chub is convex in shape and bends outwards.

The dorsal fin of the dace is concave in shape and bends inwards.

## where to find them

Chub are present throughout Europe, except Iceland, Ireland, southern Spain and southern Italy.

They are usually found in running water, but they can be caught in stillwaters, when they are often escapees from flooded rivers. Look for steady-flowing lowland or middle reaches of a river; they don't tend to thrive in the faster-flowing upper reaches, which are inhabited by trout, grayling and salmon.

## what do they feed on?

Chub are omnivorous. They eat small fish, insects and worms, as well as weed and berries. Small specimens tend to feed on worms and fry, while those over 0.9kg (2lb) will target minnows, roach and dace.

Although they don't have teeth in their mouths, they do have powerful pharyngeal teeth at the back of their throats, and these are capable of crushing almost any food item. Even crayfish, with their hard shells, don't last long in a chub's mouth.

## how to fish for them

River chub can be approached by floatfishing or legering. Many anglers like to trot for chub – that is, to allow a float to be taken by the natural current of a river with a bait like maggots or a worm underneath. Other anglers prefer to cast a leger near to overhanging trees with something like bread or cheesepaste on the hook.

Stillwater chub are traditionally harder to catch than those in running water. If you want to fish for them, try floatfished maggot, but you'll need plenty of loose feed to keep them interested.

## six baits to try

| 1 | Cheesepaste | 4 | Slugs |
| 2 | Bread | 5 | Maggots |
| 3 | Worms | 6 | Boilies |

**undercut banks** Unlike lakes and ponds, many rivers have undercut banks. These occur when the strong current of the river eats away at the bank and provides shelter out of the main flow for a number of species, including chub. You won't be able to see them, but it's always worth putting bait right under your rod tip. Often you'll get a bite immediately!

# know your fish
# rudd *(rutilus erythrophthalmus)*

Rudd are shoal fish, and small specimens can move around in groups of more than 300. Big fish – those over 0.9kg (2lb) in weight – are beautiful but increasingly rare.

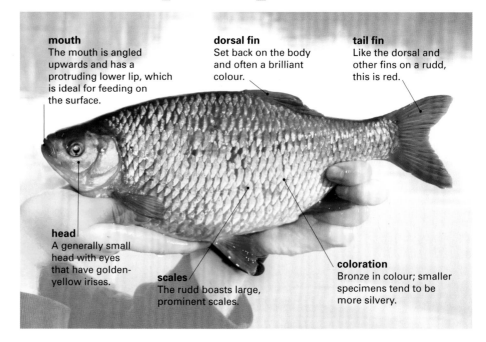

**mouth**
The mouth is angled upwards and has a protruding lower lip, which is ideal for feeding on the surface.

**dorsal fin**
Set back on the body and often a brilliant colour.

**tail fin**
Like the dorsal and other fins on a rudd, this is red.

**head**
A generally small head with eyes that have golden-yellow irises.

**scales**
The rudd boasts large, prominent scales.

**coloration**
Bronze in colour; smaller specimens tend to be more silvery.

## how to recognize them
Rudd are sometimes mistaken for roach, but they can be identified by the bright red fins and golden flanks that make them one of Europe's most beautiful fish. Other obvious differences include a steeply angled, protruding bottom lip, a set-back dorsal fin (unlike the roach) and a deep body.

They don't tend to grow big, and most of the rudd you will catch will weigh 60–115g (2–4oz). A good rudd is one over 0.45kg (1lb), with a 1.13kg (2lb 8oz) fish a genuine specimen. In parts of continental Europe, especially France, they have been known to exceed 2.27kg (5lb), but these fish are rare.

## where to find them
Rudd are widespread in France and eastwards throughout Europe, except for the far northern regions. They are mainly found in stillwaters, particularly those lined with reeds or trees, where they can feed on fallen insects. Gravel pits, ponds and canals all hold rudd, as do rivers, albeit in smaller numbers.

## what do they feed on?
The shape of the rudd's mouth gives away where it likes to feed most – off the surface – and in warm weather they love to cruise in the top layers of water, snatching insects, although they also feed on crustaceans.

## how to fish for them
Floatfishing is without doubt the best way to catch rudd. Warm summer days provide ideal conditions, and a light waggler, bulk-shotted around the base, will allow the bait to fall freely through the water. This is called fishing on the drop (see page 32), and it is a great method for catching rudd. Maggots and casters are the best forms of bait, but if you decide to fish with a bait on the bottom, bread and sweetcorn will be good alternatives.

## six baits to try
1  Maggots
2  Casters
3  Bread
4  Sweetcorn
5  Lobworms
6  Floating crust

# know your fish
# crucian carp *(carassius carassius)*

Although this small fish belongs to the same family as the common carp, it is more closely related to the common goldfish. It is widely regarded as one of Europe's prettiest freshwater fish.

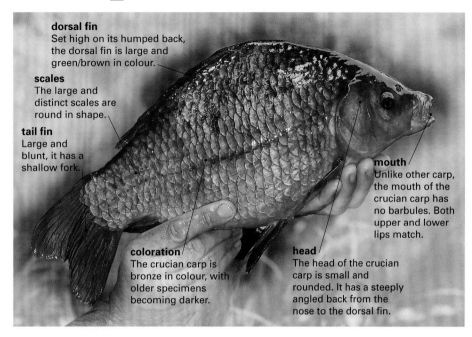

**dorsal fin**
Set high on its humped back, the dorsal fin is large and green/brown in colour.

**scales**
The large and distinct scales are round in shape.

**tail fin**
Large and blunt, it has a shallow fork.

**mouth**
Unlike other carp, the mouth of the crucian carp has no barbules. Both upper and lower lips match.

**coloration**
The crucian carp is bronze in colour, with older specimens becoming darker.

**head**
The head of the crucian carp is small and rounded. It has a steeply angled back from the nose to the dorsal fin.

## how to recognize them

The crucian carp differs from common, leather and mirror carp not only in being much smaller but also in not having any barbules around its mouth. It has a deep, golden-brown body. These fish rarely exceed 1.8kg (4lb) in weight, and you should regard anything over 0.45kg (1lb) as a good fish, and one weighing 0.9kg (2lb) as a specimen.

Unfortunately, with more and more carp being stocked into waters across Europe, a true strain of crucian carp is becoming more and more rare. The two species breed, creating a hybrid. A larger tail, more rounded body and barbules are the characteristics of a hybrid.

## where to find them

The crucian carp is native to all central and eastern European countries and has been introduced to Spain, France and Britain. They love small, weedy ponds but will also, if less often, be found in canals and slow-moving rivers. They are a hardy fish and can survive in water with low oxygen levels that would normally kill most other species.

## what do they feed on?

They like to forage for food on the bottom and will routinely eat bloodworms, midge larvae and snails.

## how to fish for them

Floatfishing is the best way to catch crucian carp. They are known as shy-biting fish, so legering is not really appropriate. A finely shotted float, with just a few millimetres showing above the surface, is ideal, with bread, sweetcorn or maggots being the best bait.

You will find a pole the best way of converting bites to hooked fish because of its superior way of presenting a bait.

## six baits to try

| | |
|---|---|
| 1 Bread | 4 Sweetcorn |
| 2 Maggots | 5 Pellets |
| 3 Casters | 6 Redworms |

# know your fish
# perch (*perca fluviatilis*)

**dorsal fin**
Another striking feature of the species, the dorsal fin is large and sharply spined.

**scales**
Perch scales are small and bedded deeply within the body. They are rough to the touch.

**head**
The large eyes are yellow with black pupils.

**mouth**
Large and flexible, the mouth of the perch is capable of engulfing large food items.

**tail fin**
Short and stubby, the tail fin is bright red, just like the anal and pectoral fins.

**coloration**
The perch has a dark brown back with olive flanks and a white belly. The markings are distinctive, with four to six black bars running down each side of the fish.

## factfile

Most female perch grow much bigger than the males.

Spawning occurs between early spring and early summer, when the female lays up to 350,000 eggs in thin strands over twigs, stones or weed.

Clear water is vital to catching perch because they hunt by sight.

Perch will eat members of their own species, with bigger specimens happily feeding on smaller relatives.

If you catch a perch, handle it carefully so that it cannot raise its spiny dorsal fins, which can give a sharp prick, although, contrary to legend, the spines are not poisonous.

Perch are probably the most striking of all European species and are instantly recognizable. Small specimens hunt in shoals, whereas bigger ones tend to swim in small packs of two or three.

## how to recognize them

These handsome fish have olive-green flanks marked by six or seven black stripes and a big, spiny, double dorsal fin. The lower fins and tail are vivid red, but generally the fish are perfectly camouflaged to hide among weed and underwater snags, where they wait to ambush small fish.

Small perch, weighing 60–115g (2–4oz), are easy to catch, but the bigger ones are more difficult. A perch of 0.45kg (1lb) is a good fish, and anything above 0.9kg (2lb) is a specimen, although fish of 1.8kg (4lb) or more are reported around Europe every year, and perch of 3.17kg (7lb) have been found in some European countries, but these are rare.

## where to find them

Perch are widespread in all water systems of Europe, apart from in the far north and south. They prefer stillwaters, like gravel pits and lakes, and slow-moving rivers, but they can also be found in weir pools. Clear water is essential – perch hate colour because it restricts what they can see, and if they can't see, they can't feed.

Look for overhanging bushes, underwater snags, bridges or any other cover because perch love to hide as they wait to ambush their prey.

## what do they feed on?

Perch are predators, and their staple diet includes small fish, such as roach and rudd, but they also eat insect larvae, including bloodworms, if there are no fish. However, in these waters the perch will not grow well and will weigh only 85–115g (3–4oz).

## how to fish for them

Fish make excellent bait, and small roach or even perch, about 60g (2oz), are one of the best ways of catching them. Try using a live or dead fish under a float. Remember, always check the rules at the fishery because some venues don't allow the use of livebaits. Look for the shadiest, most overgrown area because this is where the perch will be.

A lobworm makes an excellent bait, and the float is the best method. Perch hate the resistance that a leger weight can provide. Small perch can be caught on maggots – red ones are the most effective.

You can also try spinning (see page 57). Perch love chasing their prey and often mistake the spinner for a fish and attack it.

## six baits to try

1  Small roach, rudd or perch
2  Lobworms
3  Spinners or lures
4  Maggots
5  Casters
6  Jigs

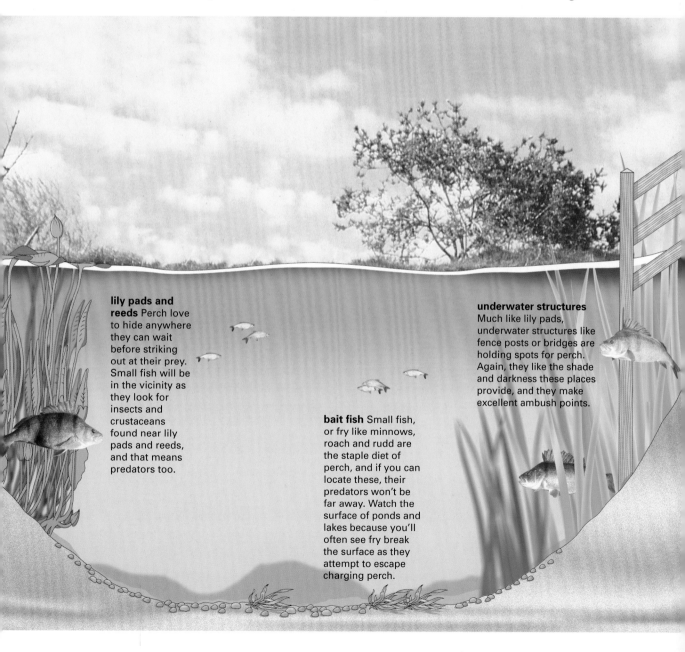

**lily pads and reeds** Perch love to hide anywhere they can wait before striking out at their prey. Small fish will be in the vicinity as they look for insects and crustaceans found near lily pads and reeds, and that means predators too.

**bait fish** Small fish, or fry like minnows, roach and rudd are the staple diet of perch, and if you can locate these, their predators won't be far away. Watch the surface of ponds and lakes because you'll often see fry break the surface as they attempt to escape charging perch.

**underwater structures** Much like lily pads, underwater structures like fence posts or bridges are holding spots for perch. Again, they like the shade and darkness these places provide, and they make excellent ambush points.

# know your fish
# pike (*esox lucius*)

**dorsal fin**
Set well back on the body, the dorsal fin is small in relation to the size of the fish.

**coloration**
The back of the pike is olive, and it has a yellow belly. Its true distinction comes in its markings, and the flanks of a pike are flecked and spotted with yellow.

**mouth**
The powerful jaws are lined with hundreds of tiny, razor-sharp teeth.

**head**
The head is huge, and the eyes of a pike are set high on its head to help with vision.

**scales**
The scales on a pike are small and set deeply into the fish's body.

**tail fin**
The tail fin, together with the dorsal and anal fins, are close together, and this helps the pike accelerate at great speed.

factfile

The pike has hundreds of razor-sharp teeth in its mouth.

Female pike are much larger than males and on average carry about 20,000 eggs, although fish over 9kg (20lb) can carry many more.

Although they feed on fish, small mammals and waterfowl, no pike has ever been known to attack a human.

Despite their fearsome reputation, pike are delicate creatures and great care should be taken in returning them to the water.

In some European countries the pike is considered a delicacy. It is said to have a 'muddy' taste.

The pike is a fearsome-looking creature, which has rightly earned its reputation as the king of the underwater world. These fish are solitary hunters with exceptional eyesight, and in murky and coloured water they rely on their sense of smell, which is also excellent.

## how to recognize them

With its green and yellow mottled markings, its long and lean body shape and pointed head, full of tiny, razor-sharp teeth, the pike is perfectly built to hunt other fish.

They grow big, with pike up to 36kg (80lb) having been reported from European waters. Any fish over 4.54kg (10lb) is noteworthy, and one over 9kg (20lb) should be regarded as a specimen. Catch a pike of 13.6kg (30lb) and it is truly a fish of a lifetime.

## where to find them

Pike are naturally found in watercourses from the Pyrenees eastwards through most of Europe. They will be found wherever there is a good supply of fish for them to eat, from lakes, ponds and slow-moving rivers to drains and gravel pits.

At each of these venues they will probably be the biggest freshwater fish present, growing fat on species such as roach, rudd and perch. Choosing a swim can be difficult because pike like to roam, but look for features such as overhanging trees, reeds and underwater gravel bars, where the pike will be able to hide before striking out at their prey.

## what do they feed on?

Like perch, pike feed on other fish and are not averse to attacking and devouring their own young. A fish in excess of 4.54kg (10lb) is quite capable of eating a fish of about 0.9kg (2lb). In addition to fish, they eat frogs, small mammals and even waterfowl – and this is probably why they have earned their fearsome reputation.

Pike have set feeding patterns, and they feed hard after spawning, which normally takes place in early to mid-spring.

They are regarded as a winter species, often feeding when nothing else will, but you will still need to target particular parts of the day, normally first and last light. Summer pike fishing can also be prolific, but not only are the fish smaller at this time of year, they are also more vulnerable to low oxygen levels in the water.

## how to fish for them

Pike are usually caught on dead or livebaits. While they can be taken on roach, rudd and small bream, they will also eat seafish, like lamprey, smelt and mackerel. These can be presented under a float or legered, but you will need strong tackle to catch these hard-fighting fish.

Lure fishing can be successful. Like perch, pike hunt by sight, and a carefully worked lure, made to look like a real fish, can work wonders.

## six baits to try

1  Roach
2  Skimmer bream
3  Mackerel
4  Lamprey
5  Lures
6  Smelt

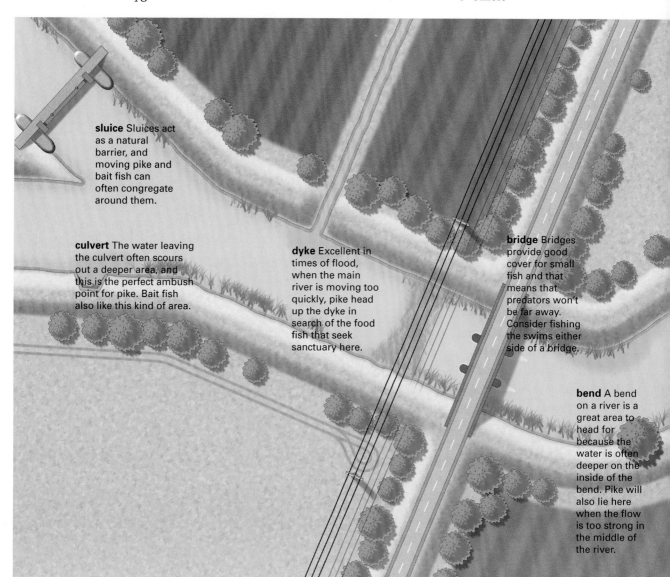

**sluice** Sluices act as a natural barrier, and moving pike and bait fish can often congregate around them.

**culvert** The water leaving the culvert often scours out a deeper area, and this is the perfect ambush point for pike. Bait fish also like this kind of area.

**dyke** Excellent in times of flood, when the main river is moving too quickly, pike head up the dyke in search of the food fish that seek sanctuary here.

**bridge** Bridges provide good cover for small fish and that means that predators won't be far away. Consider fishing the swims either side of a bridge.

**bend** A bend on a river is a great area to head for because the water is often deeper on the inside of the bend. Pike will also lie here when the flow is too strong in the middle of the river.

# know your fish
# roach (*rutilus rutilus*)

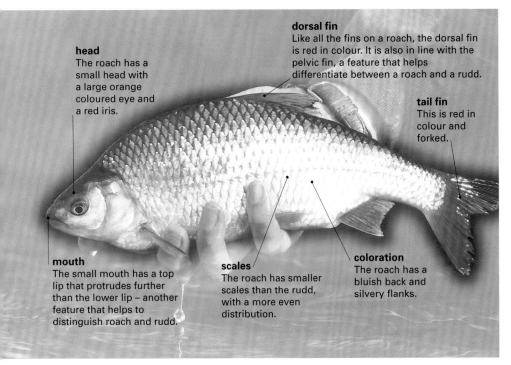

**head**
The roach has a small head with a large orange coloured eye and a red iris.

**dorsal fin**
Like all the fins on a roach, the dorsal fin is red in colour. It is also in line with the pelvic fin, a feature that helps differentiate between a roach and a rudd.

**tail fin**
This is red in colour and forked.

**mouth**
The small mouth has a top lip that protrudes further than the lower lip – another feature that helps to distinguish roach and rudd.

**scales**
The roach has smaller scales than the rudd, with a more even distribution.

**coloration**
The roach has a bluish back and silvery flanks.

factfile

Roach are the most widespread of coarse fish and are found in almost all watercourses.

They are a shoal fish, often moving in groups of up to 3,000 fish.

Roach can live for up to 12 years.

In their natural environment they eat all sorts of small invertebrates, but bloodworms are their favourite food.

Big roach are rare, and one of over 0.9kg (2lb) is considered a genuine specimen.

The roach is not known as the angler's friend for nothing. They are widespread throughout nearly all watercourses and can be found in vast shoals. Small fish are never shy of taking a bait, and many fishermen will find that the roach is the first fish they catch.

## how to recognize them
Often confused with rudd, the roach has a largely silvery body with an orange eye and brilliant red fins. The most notable difference between the two species is their mouths: roach have level lips; rudd have protruding bottom lips.

Roach don't grow to a huge size. Catch one over 0.45kg (1lb) and you have a fish to be proud of, and one of 0.9kg (2lb) is a real specimen. Although they have been known to grow to more than 1.8kg (4lb), such specimens are rare, and most waters hold a huge head of shoal fish of 60–170g (2–6oz).

They can become stunted in waters where there are vast numbers, all of the same small size and all competing for limited food supplies. Where this happens, anglers can catch hundreds of fish, all weighing 60–115g (2–4oz).

## where to find them
The roach is one of the commonest fish in Europe, being found throughout Britain and France and east into Russia. They occur in ponds, lakes, gravel pits, canals and rivers, and although they prefer stillwater and slow-moving rivers, they are a tolerant species and will often survive pollution better than other species.

Spawning takes place in the shallows between mid-spring and early summer, with the average female fish laying about 20,000 eggs per 0.45kg (1lb) of body weight.

## what do they feed on?

In their natural environment, roach feed on the bottom, rooting around for food such as snails and insect larvae, with bloodworms being a favourite. Although they are mainly bottom-feeders, they will come to the surface when the water is warm to take insects, and they will often reveal their presence by rolling.

## how to fish for them

Roach love hempseed, which is an excellent loose feed, keeping fish in the swim without filling them up. It can be used as hookbait, although its tiny size makes it difficult to hook. An alternative is to feed hempseed and use a maggot on the hook. A float is the best method of presentation, with light lines and small hooks being the order of the day.

## six baits to try

1 Hempseed    4 Bread
2 Maggots    5 Worms
3 Casters    6 Mini-boilies

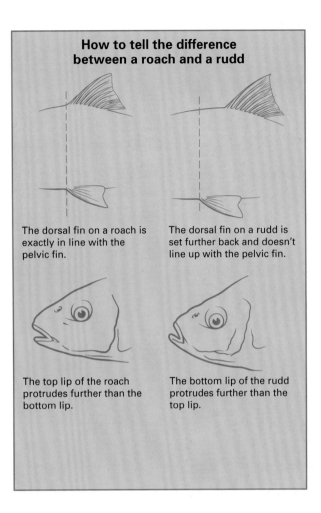

**How to tell the difference between a roach and a rudd**

The dorsal fin on a roach is exactly in line with the pelvic fin.

The dorsal fin on a rudd is set further back and doesn't line up with the pelvic fin.

The top lip of the roach protrudes further than the bottom lip.

The bottom lip of the rudd protrudes further than the top lip.

## hybrids

The word hybrid is used to describe a fish of mixed parentage that carries physical traits of two species. Roach, rudd, bream, carp, crucian carp and chub usually spawn in shoals, and once the females have laid their eggs, the males release a cloud of sperm to fertilize them. Consequently, it is not uncommon for the eggs to be fertilized by a different species of fish. The most common forms of hybrid to be found are roach and rudd (see right); bream and roach; common carp and crucian carp; and chub and roach.

# know your fish
# eel *(anguilla anguilla)*

The freshwater eel is unmistakable and can best be described as resembling a snake. They are unique among coarse fish in spending part of their lifecycle in the sea. Big fish – those over 0.9kg (2lb) – can already be up to 20 years of age.

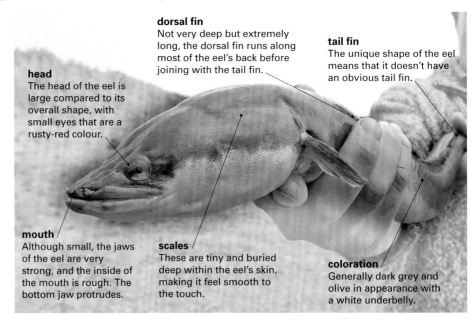

**dorsal fin**
Not very deep but extremely long, the dorsal fin runs along most of the eel's back before joining with the tail fin.

**tail fin**
The unique shape of the eel means that it doesn't have an obvious tail fin.

**head**
The head of the eel is large compared to its overall shape, with small eyes that are a rusty-red colour.

**mouth**
Although small, the jaws of the eel are very strong, and the inside of the mouth is rough. The bottom jaw protrudes.

**scales**
These are tiny and buried deep within the eel's skin, making it feel smooth to the touch.

**coloration**
Generally dark grey and olive in appearance with a white underbelly.

## how to recognize them

Long and extremely lean, eels have olive-brown flanks and deeply buried scales, which make their bodies smooth and almost silk-like to touch. They have a long dorsal fin and two tiny pectoral fins. They are becoming increasingly rare across all of Europe, and big ones are scarcer still. A fish over 0.9kg (2lb) can be considered big, especially when you consider that they grow about 0.45kg (1lb) every ten years. Anything over 1.8kg (4lb) is a real specimen.

## where to find them

Eels are widespread throughout Europe. They are born in the Sargasso Sea, in the north Atlantic, and reach European shores as tiny elvers no more than a few centimetres (inches) long. They penetrate river systems and from there migrate to any pool, lake, canal or gravel pit they can reach. They are known to cross land in order to get to water. Adult eels attempt to return to the Sargasso Sea to spawn, but many remain land-locked and are unable to escape. These fish tend to grow the biggest.

## what do they feed on?

Eels are largely nocturnal feeders and love to scavenge on dead and dying fish. They will also eat the corpses of any other animal matter, as well as invertebrates.

## how to fish for them

Eels eat virtually anything live, so almost any method will work. However, they have a particular love of maggots and worms, so these are ideal baits.

Legering after dark is the best way to catch big eels. Try three or four lobworms – break them to release the enticing juices into the water – in the margins of a lake. They also love small dead fish too, so an 8cm (3in) roach is a great bait. If you are going to use dead fish, try throwing in some mashed-up fish around your bait. This is a great way of encouraging eels into the area.

## six baits to try

1 Lobworms
2 Dead roach
3 Dead rudd
4 Maggots
5 Casters
6 Luncheon meat

# know your fish
# zander *(sitzostedion lucioperca)*

Often believed to be a cross between a pike and a perch, and sometimes called a pikeperch, the zander is in fact a predatory species in its own right, although it belongs to the same family as the perch.

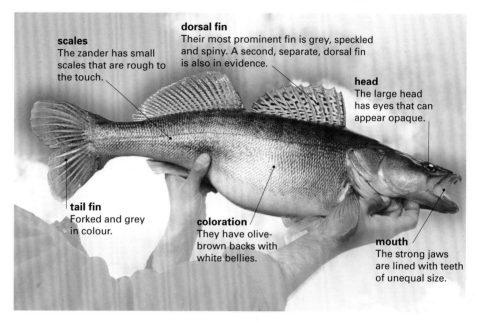

**scales**
The zander has small scales that are rough to the touch.

**dorsal fin**
Their most prominent fin is grey, speckled and spiny. A second, separate, dorsal fin is also in evidence.

**head**
The large head has eyes that can appear opaque.

**tail fin**
Forked and grey in colour.

**coloration**
They have olive-brown backs with white bellies.

**mouth**
The strong jaws are lined with teeth of unequal size.

## how to recognize them

Zander have light olive-brown backs with grey bars down the flanks and two dorsal fins, the first of which is long and spiny. They also have large heads full of tiny, razor-sharp teeth. Their eyes often appear opaque because they have adapted to feed in murky water.

In several European countries, fish weighing more than 9kg (20lb) have been found, although these are becoming increasingly rare, but anything over 3.63kg (8lb) should be considered a good catch, with a double-figure fish a very rare specimen.

## where to find them

Zander are found throughout northeastern France and central Europe and east into Russia. They have also been introduced to Britain, but their introduction is sometimes regarded with suspicion and they are blamed for poor fish stocks. They frequent lakes, canals and, especially, drains and slow-moving rivers. Some stillwaters have stocked them and in such places they can grow to 4.54kg (10lb) quickly.

## what do they feed on?

These fish are predators and like to feed on other species. Unlike pike and perch, they can feed in murky water and will do so on and off all day. In clearer water they tend to be nocturnal hunters and prey on all manner of smaller fish at night.

## how to fish for them

Most zander are caught on floatfished or legered deadbaits. They have a particular preference for small coarse fish, no bigger than 8–10cm (3–4in) long, with roach, rudd and skimmer bream ideal. Like perch, they don't appreciate any form of resistance, so your rig must be free-running.

Lures provide another method of capture, but only use these in clear water – if the water is murky, the fish won't be able to see your lure and you could wait some time for a bite.

## six baits to try

1 Dead roach
2 Dead rudd
3 Gudgeon livebait
4 Lures
5 Lamprey section
6 Lobworms

# know your fish
# tench (*tinca tinca*)

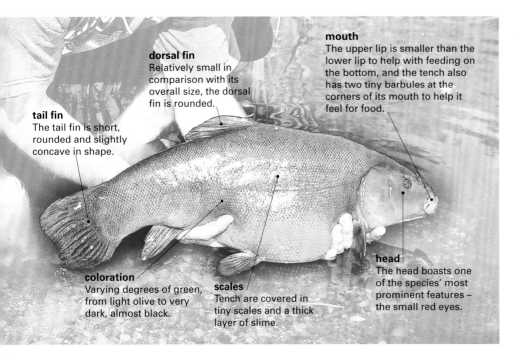

**dorsal fin**
Relatively small in comparison with its overall size, the dorsal fin is rounded.

**mouth**
The upper lip is smaller than the lower lip to help with feeding on the bottom, and the tench also has two tiny barbules at the corners of its mouth to help it feel for food.

**tail fin**
The tail fin is short, rounded and slightly concave in shape.

**coloration**
Varying degrees of green, from light olive to very dark, almost black.

**scales**
Tench are covered in tiny scales and a thick layer of slime.

**head**
The head boasts one of the species' most prominent features – the small red eyes.

factfile

Tench are known as the doctor fish, and legend tells that the slime that coats the body of the tench has healing properties.

Tench can be found in most lowland lakes, some slow-moving rivers and canals.

They thrive best in weedy waters with a rich, muddy bottom, although they are quite tolerant of stagnant conditions.

The average tench lives to 10–15 years of age.

In France and Germany tench is regularly eaten, and it is said to have a strong taste.

The tench is one of the most recognizable of all freshwater fish with its distinctive green colouring, velvety skin and tiny red eyes. For its willingness to fight – pound for pound it is rated by many as one of the hardest-fighting specimens – and its appearance, the tench is one of the angler's favourite species.

## how to recognize them

Tench can range in colour from almost black to pale silver-green, and there is a rare ornamental variety, the golden tench, which is golden-yellow with black markings.

Any tench over 1.36kg (3lb) is a good fish, with a specimen recognized as being anything over 3.63kg (8lb). Fish of 4.54kg (10lb) are rare, but the species is known to grow over 6.8kg (15lb).

## where to find them

They are found throughout much of Europe, except northern Scotland, Iceland and Russia. Anglers wishing to target tench should opt for a pond, lake or gravel pit. Although they do still populate canals and even rivers, they are much better suited to rich stillwaters where they can feed in the silt and mud on the bottom.

They do not spawn unless the water temperature has warmed up sufficiently, and that means they lay eggs later than most coarse fish. Many specimen anglers catch the biggest tench in late spring or early summer – just before spawning – and some fish can weigh an extra 0.9–1.36kg (2–3lb) if caught when spawn-bound. Big females can carry close to one million eggs. Each pair takes several weeks to spawn, leaving clusters of eggs stuck to the stems of water plants. May or June are the two most popular months for spawning.

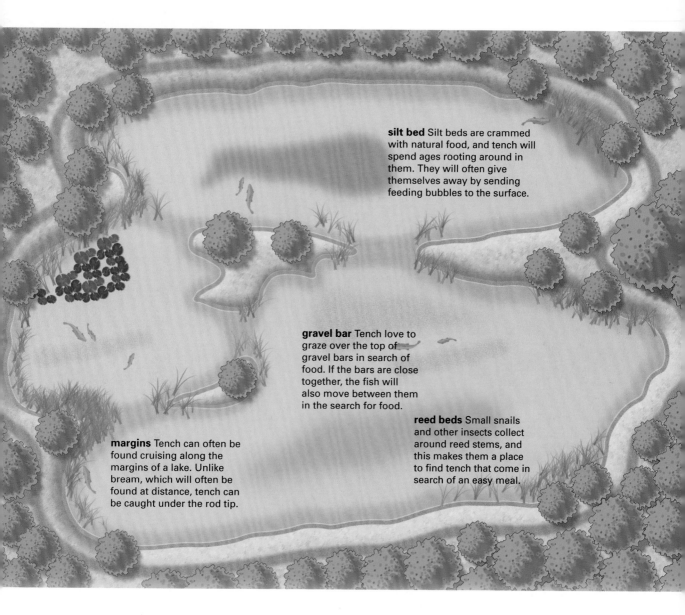

**silt bed** Silt beds are crammed with natural food, and tench will spend ages rooting around in them. They will often give themselves away by sending feeding bubbles to the surface.

**gravel bar** Tench love to graze over the top of gravel bars in search of food. If the bars are close together, the fish will also move between them in the search for food.

**reed beds** Small snails and other insects collect around reed stems, and this makes them a place to find tench that come in search of an easy meal.

**margins** Tench can often be found cruising along the margins of a lake. Unlike bream, which will often be found at distance, tench can be caught under the rod tip.

## what do they feed on?

Tench are almost exclusively bottom-feeders, rooting around in the mud for worms, snails, mussels and bloodworms. When they feed, they release tiny pockets of bubbles from their mouths – a signal that is sure to set the angler's heart racing when it appears next to his float!

They prefer to feed at dawn, although tench will feed through the day with another hotspot at dusk. They are not known as night feeders.

## how to fish for them

Your approach will depend on what venue you are fishing. If you are targeting an estate lake or canal, floatfishing is definitely the best method, with maggots, bread, sweetcorn or a lobworm on the hook.

If you are on a gravel pit, the fish will tend to be further out, making legering the best tactic. Here, use a leger bomb or a swimfeeder to get that extra distance, and small boilies tend to be most effective on the hook. Tench respond very well to pre-baiting – putting groundbait into the swim the day before you intend to fish.

## six baits to try

1 Lobworms
2 Sweetcorn
3 Red maggots
4 Boilies
5 Luncheon meat
6 Bread

# other common species

Several other fish are caught in European waters, even if on a less-frequent basis. This can be because they are rare or because they are too small to be considered a target for anglers.

**Gudgeon** *Gudgeon might look like a smaller version of barbel, but they are a species in their own right.*

## gudgeon (*gobio gobio*)

Inexperienced anglers often mistake gudgeon for small barbel, but this is a different species that grows to just 225g (8oz). The barbel has two sets of barbules, but the gudgeon has only one set, and its coloration – mostly silver with round, dark blotches along the flanks – is different.

It is found throughout Europe and lives mainly in running water, although also found in canals.

**Ide** *Ide might resemble roach or even chub, but they are far rarer than either species.*

**Bleak** *Bleak are a tiny river species that live in huge shoals and are often considered a nuisance by anglers.*

## ide (*leuciscus idus*)

The ide, sometimes known as orfe, is present throughout Germany, Holland, Scandinavia and most European countries, where it can be found in lakes and rivers. Ide are often mistaken for chub or even roach, but they are far rarer than either of those species. They love to eat caddis larvae, nymphs, snails and shrimps, and grow, on average, to 1.8kg (4lb).

## bleak (*alburnus alburnus*)

This tiny species, which lives in vast shoals, is significant to anglers for only one reason: it can be easily caught. Almost always found in rivers, it is widespread throughout nearly all of Europe's watercourses. It has a pale, silvery coloration with large eyes. Bleaks feed on nymphs and plankton in midwater and rarely grow longer than 12–15cm (5–6in) or weigh more than 85–115g (3–4oz).

**Grayling** *Grayling are a river species that are characterized by their huge dorsal fin.*

**Minnow** *Minnows are another tiny fish that will readily take an angler's bait. Again, somewhat of a nuisance species.*

## grayling (*thymallus thymallus*)

Although the grayling is considered a game fish, it is present in waters throughout England and Wales, France and central Europe. Exclusively a river species, it can be recognized by the big, sail-like dorsal fin and brilliant silver coloration. The fish rarely grow in excess of 1.8kg (4lb), and feed both on the bottom and on the surface.

## minnow (*phoxinus phoxinus*)

Although not necessarily the quarry of the angler, the minnow can be the first fish many encounter, such are their numbers. They are only a tiny fish, growing to just 10cm (4in) long, but they are widespread throughout Europe. They have small grey fins, with light olive-brown sides, which are dappled with characteristic dark blotches.

**Nase** *Nase are found in several European countries, but not the UK. They are recognizable by their blunt, bulbous snout.*

**Grass carp** *Grass carp can grow to huge sizes and have become an increasingly popular species.*

## nase (*chondrostoma nasus*)

Found mainly in river systems in France, Russia and central European countries, the nase is recognizable by its odd-shaped mouth, which is underslung, small and positioned under a blunt, bulbous snout. It feeds on algae, which it scrapes from rocks and stones. It has a long, narrow body with reddish fins and silvery-grey coloration.

## grass carp (*ctenopharyngodon idella*)

Although a member of the carp family, the grass carp looks nothing like a common, mirror or leather carp. They still grow large – in excess of 18kg (40lb) – but have a different shape and lack the barbules of their cousins. They have long, lean bodies and silvery coloration.

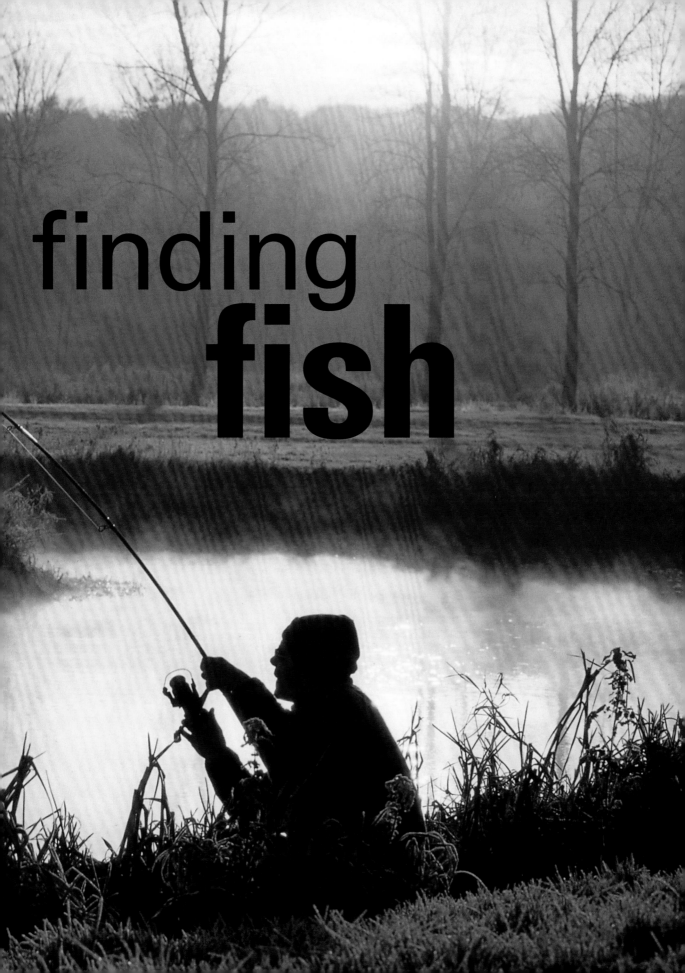

# finding
# fish

# finding fish
# in lakes and ponds

*Left A classic lake, fringed with reeds and covered with lillies, is the perfect place to catch fish.*

## fish species

**The common species in lakes and ponds include:**

- Roach
- Rudd
- Tench
- Bream
- Carp
- Pike
- Perch

Although some lakes are completely natural, having been created by ancient land shifts, many are the result of agricultural practices. Farmers' land is littered with ponds and lakes, originally built to water livestock or for irrigation, and these have become havens for coarse fish. Many of these waters, most of which are more than 50 years old, are now used for a second purpose – angling.

## a variety of species

Roach and rudd will often be found in great numbers, but they are, as a result, unlikely to reach specimen size. However, because there are so many of them, they are easy to catch. Tench and bream will feed over silt beds and give themselves away by blowing feeding bubbles to the surface. Most established lakes and ponds will be surrounded by overhanging trees, and this is where carp can be found.

The predators to be found in such waters are pike and perch. Both love to find shady areas where they can wait in ambush for small fish.

## a rich environment

Because of their age, these lakes and ponds tend to be well established, not just by fish but also by all manner of wildlife, both above and below the water's surface. As such, they can be thoroughly enjoyable places to fish. The bottoms of these venues are usually rich in bloodworms and larvae, while water beetles, leeches, snails and water boatmen give the fish plenty to feed on.

This environment encourages prolific weed growth, with milfoil, hornwort and all manner of pondweed prevalent. Expect to see lily pads and reedmace too.

## how do I fish a lake?

Most lakes and ponds tend to have a uniform shape, shelving gently down from the margins to depths of 2–2.4m (6–8ft). They also tend to be mature, surrounded by trees and vegetation and with heavy weed growth.

Lakes can be fished in a variety of ways. You can floatfish with a rod or pole for roach, rudd and perch, or use a swimfeeder to catch bream. Carp and tench can also be caught from lakes using standard leger tactics. Older lakes can have silty bottoms, so keep this in mind when fishing.

**Tench**
Look for the margins if you want to find tench. These fish will come remarkably close to the bank in search of food items, especially at dawn or dusk. You will be able to spot tench by looking for tiny bubbles – a sign that fish are feeding.

**Pike**
The pike is almost always the top predator in any lake it inhabits. Camouflaged by its mottled green coloration, it will be found in submerged reeds or in snags waiting patiently before striking with great speed at its prey.

**Perch**
As predators, perch will look for dark, dingy areas of lakes where they can wait in ambush for small fish like roach. Look for underwater snags such as sunken trees or platforms and perch won't be far away.

**Rudd**
With its upturned mouth, the rudd likes nothing better than to pick small flies off the surface. Almost exclusively a summer fish, it used to be prevalent in most stillwaters across Europe, but due to increasingly poor water conditions – and competition from roach – it is far less common today.

**Carp**
These fish love to spend time feeding over silt or gravel patches and will often be found near cover like overhanging trees. They are more often than not the biggest inhabitants of any lake and will bully other fish out of the way.

**Bream**
Another shoal fish that spends most of its time feeding over silt beds in deeper areas of lakes and ponds. Tell-tale signs of bream in a fishery are distinctive feeding bubbles and rolling fish. Dawn and dusk are the best times to spot them. Bream are almost exclusively bottom-feeders and are caught with baits presented on the bottom.

**Roach**
Small roach will be found in large shoals, and their numbers give them relative safety from predators. They like to graze over gravel beds, feeding on bloodworms and daphnia, but in many stillwaters they grow to relatively small sizes because of the numbers competing for food.

# finding fish
# in canals

*Left Canals, a small fish haven, provide a great spot for anglers.*

Canals were originally dug to allow barges to transport goods from one city to another in the 1700s. With this purpose long since redundant, they are now mostly used for leisure pursuits. There are literally hundreds of canals across Europe, many of which run for great distances, and nearly all hold a good head of coarse fish. Some areas on canals are better than others, but expect to find fish throughout entire stretches.

## fish species

**The common species in canals include:**

• **Roach**

• **Rudd**

• **Bream**

• **Tench**

• **Carp**

• **Pike**

• **Perch**

## a uniform shape

Fish don't grow as big in canals as they do in lakes and rivers, largely because less food is available, but they still provide a unique challenge.

Nearly all canals are a uniform U-shape, which is a throwback to their original design for allowing barges to transport. This makes it easier to predict where the various species live.

Roach, rudd, bream and perch will be found down the central track of the canal, where they congregate in big shoals. Look to the far bank, where there is often cover, to find chub and carp, while pike will often be found lurking in the margins where they wait in pursuit of prey.

## try the opposite bank

Most canals have at least one towpath running alongside them for walkers and cyclists. More often than not, however, the far bank will be inaccessible and overgrown, providing the angler with a perfect area to target.

If the canal is still used frequently by leisure boaters, weed growth in the central track is likely to be minimal, with most of the pondweed and milfoil to be found in the margins of both the near and far bank. The bottom of these venues tends to be rich in life, with bloodworms predominant. Another staple diet of the fish will be beetles, leeches and snails.

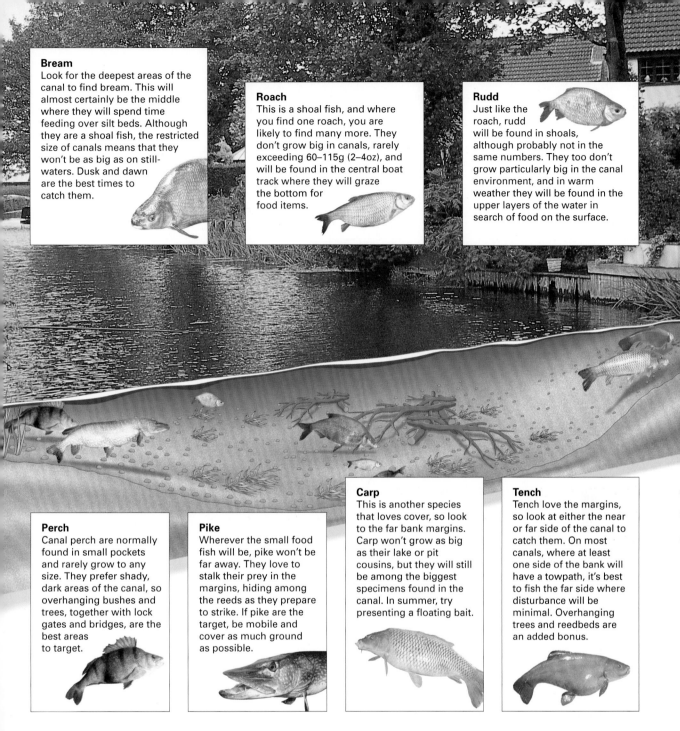

**Bream**
Look for the deepest areas of the canal to find bream. This will almost certainly be the middle where they will spend time feeding over silt beds. Although they are a shoal fish, the restricted size of canals means that they won't be as big as on still-waters. Dusk and dawn are the best times to catch them.

**Roach**
This is a shoal fish, and where you find one roach, you are likely to find many more. They don't grow big in canals, rarely exceeding 60–115g (2–4oz), and will be found in the central boat track where they will graze the bottom for food items.

**Rudd**
Just like the roach, rudd will be found in shoals, although probably not in the same numbers. They too don't grow particularly big in the canal environment, and in warm weather they will be found in the upper layers of the water in search of food on the surface.

**Perch**
Canal perch are normally found in small pockets and rarely grow to any size. They prefer shady, dark areas of the canal, so overhanging bushes and trees, together with lock gates and bridges, are the best areas to target.

**Pike**
Wherever the small food fish will be, pike won't be far away. They love to stalk their prey in the margins, hiding among the reeds as they prepare to strike. If pike are the target, be mobile and cover as much ground as possible.

**Carp**
This is another species that loves cover, so look to the far bank margins. Carp won't grow as big as their lake or pit cousins, but they will still be among the biggest specimens found in the canal. In summer, try presenting a floating bait.

**Tench**
Tench love the margins, so look at either the near or far side of the canal to catch them. On most canals, where at least one side of the bank will have a towpath, it's best to fish the far side where disturbance will be minimal. Overhanging trees and reedbeds are an added bonus.

## how do I fish a canal?

Floatfishing is usually considered the best way of getting the most from a canal, with many anglers preferring to pole fish. Poles are able to reach the far bank on most of these venues, and their near-perfect presentation means that they are a great success.

Roach, rudd, perch, bream and tench can all be caught on the pole, but if you want to target carp or pike, a specialist approach with rod and line is advisable. If carp are the target, a standard leger set-up will be the best approach with a big bait like lobworm, breadflake or a boilie.

Carp don't grow very big in a canal – a fish of 9kg (20lb) is about the maximum – but you will need to step up the strength of your tackle.

Target overhanging trees, moored boats and bays built for allowing barges to turn.

# finding fish
# in rivers

**Left** *The running waters of rivers require quite a different approach to stillwater fishing.*

## fish species

**The common species in rivers include:**

- Roach
- Chub
- Barbel
- Bream
- Perch
- Pike
- Grayling
- Dace

Rivers offer some of the best fishing there is to be had across Europe, but they require a different approach from that used for stillwater angling. Rivers can be unpredictable, and range from small, fast-flowing, shallow streams to deep, wide and sluggish waterways.

## full of life

All rivers start as a stream on high ground, but the streams grow, and as they wind through hills and mountains, many develop into big rivers, some to 100m (330ft) wide.

Rivers are usually full of life, both under the water and above it. All fish species inhabit rivers, with chub, barbel, grayling and roach among the most popular species. On the bottom, among the gravel, sand and silt, ranunculus, bulrushes and reedmace grow in abundance.

The high quality of the water in most rivers means that water lice, midge larvae, snails, bloodworms, mussels and even crayfish thrive in many places, providing the resident species with a huge larder.

## where do I find the fish?

Rivers can be difficult to 'read' for the beginner, but there are several hotspots that will almost certainly contain fish.

Fish like to congregate in well-oxygenated water, especially in warmer months, so look for slightly deeper holes at the end of fast runs, where fish will be waiting for food to be carried down into the current.

**Chub**
Chub love cover and can be extremely shy, especially in bright conditions. Overhanging trees and deep holes are perfect chub-holding areas, and it may take a loose feed like hempseed to draw them from their sanctuary before you present a bigger bait over the top.

**Bream**
Bream aren't naturally designed for very fast water (unlike the streamlined barbel) and will therefore tend to be in deeper, more sluggish water. Like their stillwater cousins, they are shoal fish, and where one is found, many more will be in attendance.

**Roach**
Small roach are normally found in quite large shoals, and they dart in and out of weedbeds in the main current, picking off passing food items. Bigger roach, those over 0.45kg (1lb), tend to be solitary and will be with just a handful of other fish.

**Pike**
Pike will be wherever there is a food source, so find the smaller fish and you'll find this fearsome predator. They will lurk under near or far bank cover, or in deeper depressions in the river-bed, as they wait to strike out at roach and dace.

**Grayling**
These can be found in remarkably shallow water where they use their streamlined bodies to move in and out of the weed to take passing food. The water needs to be well oxygenated though, so look for areas near weir pools or fast straights.

**Perch**
Cover is again the key word when it comes to perch. As predators, they will hide under near bank trees, among roots or in the shadow of bridges as they wait to strike their prey. They are very sensitive to light and water colour, and in bright sunlight or in cloudy coloured water they are difficult to catch.

**Dace**
These small fish can often be found in vast numbers, and they compete with each other for food. They don't like water that's too fast or too turbulent, preferring instead steady-paced runs or slow slacks.

**Barbel**
Fast water is the first place to look for barbel, which always prefer to be in well-oxygenated water. Again, marginal or mid-river cover is worth looking for, as well as undercut banks. Barbel will wait here in search of passing food items.

They also love the cover of weed and will cruise in and out of the cover in search of food, so look for clear areas, particularly clear gravel, because many river fish like to feed over a hard bottom.

Undercut banks (see page 69), nearby trees and reedbeds are also holding areas, with barbel, chub, perch, roach, bream, dace, pike and grayling the most likely species to be found in running water.

## how do I fish a typical river swim?

There are two basic ways of tackling a typical mid-river swim: with a simple leger set-up or with a float. Legering is probably the more effective way of catching barbel and chub, whereas the float will be more successful if roach and perch are the target.

Be prepared to be mobile, wandering from swim to swim until you find the fish. Often you will catch only a few fish before you have to move on.

A basic paternoster set-up (see pages 34–5) using luncheon meat, bread or worms will account for chub and barbel, whereas maggots, casters or punch will be most effective for roach, dace or grayling.

# finding fish
# in weir pools

**Left** *Weir pools contain all manner of species and are great in warmer weather.*

## fish species

**The common species in weir pools include:**

- Barbel
- Chub
- Bream
- Roach
- Tench
- Carp
- Pike
- Perch

Weir pools are magnets for a host of species, and they come into their own during the summer. Built across rivers to raise the upstream water level or control the downstream flow, the water that tumbles over the weir mixes with air as it falls and so becomes well oxygenated.

## fast water

The real beauty of weir pools is their design. Not only do they boast fast water, they also contain deep slacks, back eddies, slower water and shallows – basically nearly all the different river environments in one place. As such, they are home to lots of species, each of which prefers its own particular area. Remember that the riverbed in weir pools is likely to be littered with all manner of debris brought down from upstream, and this is one of the reasons fish are attracted to them.

Another feature of most weir pools is that they are often deeply undercut, which means that the concrete of the base of the weir protrudes for some distance, providing superb cover for species like barbel and chub.

## where do I find the fish?

Barbel and chub can be found in the fast water that immediately falls over the weir or even beneath the sill. Both species love oxygenated water, and in the summer months this is where you will find them.

Species like bream, tench, roach and even carp, however, prefer a much more sedate pace and will prefer to congregate at the tail of the pool or in the deeper, slower-moving slacks, out of the direct flow.

Pike and perch are never far away either, often sitting just out of the main flow where they wait to strike at prey.

No two weir pools are the same, so it's well worth spending time investigating the different areas that make up these superb locations.

## Roach
Look for roach at the end of the weir pool where the water is slightly slower and less turbulent. They love to feed over gravel beds and will often be found in big shoals, picking off any remaining food items being carried in the main flow.

## Barbel
This species loves fast-flowing water, so look to the areas where the water is at its quickest. The white water created directly beneath the weir is a definite holding spot, and if the weir is undercut, this too will be a place to look for barbel.

## Pike
Pike are never far away when there is an abundance of food fish to be had. Their streamlined bodies make them as much at home in rivers as they are in stillwaters, although they won't be in the main flow of a weir pool. They are more likely to be found lurking in weedbeds than in slower water.

## Perch
Weir pools provide a good hunting ground for perch. They are a species that loves cover and, with undercut banks and all manner of underwater debris, they are often found here in good numbers.

## Chub
Just like barbel, chub love to be in fast water. Occasionally you will find them just off the main flow where they wait in search of food items, but in summer you are far better basing your attack around the highly oxygenated water in the weir.

## Bream
The slacks and eddies off the main flow are where you'll find bream. Their body shape doesn't enable them to thrive in very fast water, and they are happier in deeper slacks where they graze on food brought down from upstream.

## Tench
Although predominantly a stillwater species, occasional tench will be found in weir pools. Like bream, they prefer the slower-paced water out of the main flow, so target deeper slacks and eddies.

## Carp
Carp are attracted to weir pools by the amount of food washed down by the flow. Because they won't be present in large numbers, it's difficult to target them specifically. However, should you want to try, look for slower, deeper water out of the main flow.

# how do I fish a weir pool?
If you want to catch barbel or chub, look for the fast water and use stout tackle, because both fight hard in the fast flow. Heavy leads will also be needed to hold bottom in the current, so floatfishing is not an option.

In the slower water and slacks, which is where you will find bream and tench, a swimfeeder and quivertip rod will be best.

If you want to floatfish, try trotting maggots at the tail of the weir pool where the bottom will be more uniform and the flow less turbulent.

# finding fish
# in gravel pits

**Left** *Gravel pits can be huge, but don't let this put you off – they can contain very big fish.*

## fish species

**The common species in gravel pits include:**

- **Roach**
- **Rudd**
- **Tench**
- **Bream**
- **Carp**
- **Pike**
- **Perch**
- **Catfish**

Unlike lakes and ponds, gravel pits are relatively recent. Throughout the 1960s, particularly in Britain, France and Germany, the demand for houses and motorways led to gravel workings, as vast quantities were needed for construction work. Once it had been excavated, huge holes were left, which eventually filled with water and now, with many reaching maturity, provide superb fishing.

## unpredictable depth

Gravel pits tend to be bigger than ponds and lakes, and the fish in them are often bigger, although at lower stock densities. All fish species can be found. Choosing a location on a huge expanse of water can often be the main problem, but if you spend time studying the water you'll get some idea of what lives where. Their original purpose means that the bottoms are usually uneven and unpredictable. Unlike ponds and canals, which have fairly uniform depths, the pits can be as much as 6m (20ft) deep in one place but extremely shallow just a short distance away. They also tend to boast lots of islands.

## where do I find the fish?

Coarse fish are creatures of habit and will have similar characteristics to those that live in other types of stillwater. Tench, carp and catfish can be found in the margins, roach and rudd will shoal, while bream will be in the deeper areas. Perch and pike will hunt close to their prey.

The water is usually clear, and gravel pits have their own rich eco-systems. The bottom will boast snails, bloodworms and larvae, and there is often plenty of plant life, especially pondweed. Suitable spawning areas are few and far between, however, so fish tend to be fewer in numbers but, with less competition for food, bigger in size.

**Tench**
Tench are margin-loving fish on any water. Even though there might be up to 40 hectares (100 acres) of water in front of you, tench will still be located under the rod tip. Holes in the weed at the bottom of the marginal shelf are perfect places to look for them.

**Roach**
In a gravel pit these shoal fish are much smaller in number but bigger in size than in other waters. Small pockets of big fish, usually over 0.9kg (2lb), will be found feeding over gravel bars.

**Rudd**
Gravel pits hold some of Europe's biggest rudd, and in summer they can sometimes be easy to spot. They love to feed on surface insects and will give themselves away by rolling out of the water. Find shallow bars in deeper water and rudd won't be far away.

**Catfish**
Mainly nocturnal hunters, catfish like to lurk in the deeper water during daylight hours before moving up and down the margins in darkness scavenging for food.

**Pike**
If you can find the food fish, you'll find pike. Again, thanks to the abundance of available food, pit pike can grow very big. In winter they prefer deeper water, but in summer they will move great distances to keep in touch with the big shoals of rudd, roach and bream. And don't ignore the margins.

**Bream**
Unlike tench, bream are often located at distance in pits. They love to roam in huge shoals, stopping to feed in deeper water before moving their way up gravel bars towards the surface. Spend time watching the water at dawn or dusk, when they will often be seen rolling.

**Perch**
When they get to sizes above 1.36kg (3lb), perch can be solitary creatures and difficult to locate. For that reason, spinning for them is the most effective method because of the distance covered. Any marginal cover is worth investigation, and the same goes for underwater snags or jetties.

**Carp**
Carp are at their biggest in gravel pits, thanks largely to the abundance of food available. They, like tench, can often be found in the margins as well as close to areas that shelve steeply from deep water to shallow. On sunny days they will cruise close to the surface in search of food.

## how do I fish a gravel pit?

Don't be put off by the size of the pit. Finding holes in the weed and determining the range or depth are usually the keys to successful fishing.

To find the depth of the water in front of you and where the clear areas are, try casting a lead weight, about 30g (1oz), into different areas of the swim. Count the number of seconds it takes to hit the bottom – one second roughly equates to 30cm (12in) – to establish depth and then gently reel in to determine the make-up of the bottom. If you struggle to reel in, you're in weed; if it comes in freely, you have found a clear area and this is where you should base your attack.

# essential
# skills

# essential skills
# choosing a **swim**

**Left** *Picking the right swim is crucial because you won't catch fish if there are none to be caught!*

jargon buster

**Features** The areas on stillwaters and rivers where you are most likely to find fish. Islands, bridges and snags are all good examples.

**Plumbing the depth** This allows you to find what depth of water you are fishing in. When you've done this, you can decide where to present your bait.

Knowing where to fish when you turn up at a fishery can be daunting. A lot will depend on the type of venue – whether it's stillwater or a river – but there are a few things you can do in advance to give you an edge.

The first of these is to ask advice. Your local tackle shop will be a mine of information and can give you a huge step up the ladder. Ask about what venues in your area are fishing well and what swims to head for. This can save lots of wasted time and energy.

You should also check the local and national press. Specialist fishing magazines carry 'Where to Fish' sections, which update readers on what fisheries are in form.

The time of day and the weather conditions are also important. Bright, cloudless, sunny days, when temperatures are high, are not much use for fishing. Much better are cloudy days with a wind to help stir up the bottom. Dawn and dusk are usually the best times because the fish are at their most active.

**Above** *Look for reedbeds, overhanging bushes and lily beds. These will be a magnet for fish.*

## ... on a stillwater

Although much depends on the species you are after, most fish will be held in a swim by certain features, which may be under or above water.

In addition to the natural features noted below, choosing a swim can be made easy by the fish themselves, which give away their location by rolling on the surface or sending up feeding bubbles. Tench, bream, carp, roach and rudd love to roll over areas where they are feeding, while bubbles are made when the fish root around on the bottom for food.

### overhanging trees

All species, especially carp and tench, cruise along the margins of the near and far banks, and shelter under trees.

### snags

Predators, such as perch, pike and eels, love to hide among broken tree branches and similar awaiting their prey.

### gravel bars

Bream, tench, carp and roach will cruise over deviations from the normal contours of a lake bed in search of food items. Find the bars by plumbing the depth carefully.

### clear patches in weed

Although fish hide among weed, they will break cover if a clear area appears. Identify these areas by looking with polarizing glasses or plumbing.

### islands

All fish will forage for food in the shallows around islands in gravel pits and lakes, which often also provide overhead cover in the form of trees and shrubs.

### reedbeds

Mostly found in nearside margins to the left or right of swims, many fish, but especially rudd, swim in and out of reeds in search of food. Reedbeds on the far bank are equally good holding spots.

## ... on a river

Picking a spot to fish on a river is usually easier than on a stillwater because the swims are far more obvious. Lakes tend to have a more uniform shape, but rivers are far more uneven and thus far more interesting to look at. As on stillwaters, remember that all fish love cover, so look for bushes that reach beyond the water's edge. Chub won't be far away.

### weir pools

Fish won't be far from such pools. Barbel and chub will be waiting in the shallow fast water immediately beyond the sill for food items to come down, while bream will be in the slacker, deeper areas. Pike and perch will be waiting to strike just out of the main flow.

### bridges

Perch like shady, dark places where they can wait to strike at their prey, and bridges are brilliant cover and a definite hotspot on a river.

### gravel runs

Barbel love fast water and will often emerge from the sanctuary of weedbeds to feed over shallow gravel runs.

### deep slacks

The bottom of rivers is extremely uneven and often includes deep holes where the water is much slower. Bream, tench and carp, species that prefer stillwaters, love to wallow in these areas, picking up food items that pass by.

### undercut banks

Unlike stillwaters, many rivers have undercut banks, where barbel and chub like to rest out of the main flow (see page 69).

### bends

As a river snakes through the countryside it will create lots of bends. On the inside of these bends the water will often be slow and deep, which provides a perfect haven for bream, carp, roach and tench.

# essential skills
# playing and landing fish

## jargon buster

**Playing a fish** The art of bringing a fish safely and securely from the point of hooking to the landing net.

**Backwind** To release line from the reel by allowing the reel handle, under pressure, to spin in the opposite direction to when retrieving a float or feeder.

Most small fish will not require much more effort than simply reeling in, but anything weighing more than 0.45kg (1lb) will need to be played. Landing fish with a net is a skill the beginner must master quickly.

The keys to safely banking any specimen are to keep a tight line and to have balanced tackle. For example, a match rod requires a similar match reel, with line strengths of 1.36kg (3lb) and hook sizes between 18 and 12. If you use a heavy carp rod and light breaking strain (BS) lines, the line is certain to snap.

## playing a fish

When a fish initially runs, apply pressure by holding the rod well up in the air where it can absorb any lunges. Never point the rod at the line because this will cause the line to break.

### releasing line from the reel

When you are attached to a powerful fish you, will need to allow your reel to give line. You can release line by switching on the anti-reverse, normally found underneath the reel (see pages 14–15), or by setting the clutch.

Some experienced anglers prefer to use the anti-reverse, which allows the reel handle to backwind when put under pressure, but for newcomers it's advisable to set the clutch, which you do by turning the dial at the back of the reel so that it allows line to come off the reel only under tension. Don't make it too tight, which will break the line, or too loose, which will allow the fish to take unnecessary amounts of line. The idea of the clutch is to give line only when it's absolutely essential.

### using sidestrain

If the fish is heading towards snags, lower the rod so that it's parallel with the surface. Do this in a swift movement and hold it there. This will turn the fish by knocking if off balance, and you will be able to take control again.

**pumping the fish**

When the fish reaches the end of the fight and is tiring, lower the rod (remember to keep the tension up) and then slowly bring it upwards. This can be repeated until the fish is ready for netting.

## landing a fish

A fish weighing over 225g (8oz) will need to be netted, but smaller ones can be swung to hand.

The most important thing to remember when landing a fish is never to rush or 'stab' at the fish. It's better to wait until it's completely tired so you that can gently draw it over the net. Inexperienced anglers are often tempted to scoop the fish up as quickly as possible, but all this does is make it more excitable and increase the chances of losing it.

Keep the landing net low in the water with one hand, while holding the rod high with the other. Make sure that the line remains taut. Remember, this is the point when most fish are lost, so take your time and ask for help if you need it.

**Right** *Never make a lunge with the net – instead, wait until the fish is tired and slowly draw it over the waiting landing net.*

## balanced tackle

Use the following as a rough guide to what rods, reels, line strength and hooks should be used together.

| rod | reel | bs line | hook size | fish size |
|-----|------|---------|-----------|-----------|
| Match rod | Match reel | 1.36kg (3lb) | 18–20 | to 1.36kg (3lb) |
| Quivertip rod | Match reel | 1.8kg (4lb) | 16–10 | to 2.27kg (5lb) |
| Avon-style rod | Specimen reel | 2.72kg (6lb) | 12–8 | to 3.63kg (8lb) |
| Light carp rod | Specimen reel | 3.63kg (8lb) | 10–6 | to 9kg (20lb) |
| Heavy carp/pike rod | Big specimen reel | 5.44kg (12lb) | 8–2 | to 18kg (40lb) |

# essential skills
# unhooking your catch

**Left** *Take care when returning fish and gently lower them into the water. Never simply throw them back.*

Once you have safely landed the fish, you must now carefully unhook it and release it, place it in a keepnet or weigh it. Whatever you do, you must make sure that you handle your catch correctly.

Remember that fish have a protective layer of slime covering their bodies and it's essential that this is not removed or the scales will be damaged. Many beginners use a cloth to handle fish, but this should definitely be avoided. Instead, wet your hands and hold the fish firmly but gently. The longer the fish is out of water, the greater the stress it will be under.

You will be able to unhook most fish by simply placing your fingers on the hook and carefully pulling it from the mouth. Sometimes, however, you will need specialist tools.

## using a discorger

A discorger is a small, thin piece of plastic that has a groove etched into the end that will help with deep-hooked fish. Deep hooking occurs when the fish swallows the bait, something that can happen when you don't hit a bite immediately. Perch are notorious for swallowing baits greedily.

Using the discorger will become second nature after a few attempts, but it's not simply a case of thrusting it into the fish's mouth, which can cause real damage. Hold the line heading into the mouth taut and slide the discorger down it until it

**Left** *A discorger will enable you to quickly and safely remove the hook. If the hook is deep in the throat of the fish, take extra care.*

reaches the eye of the hook. Press directly downwards until the hook moves freely and withdraw it from the mouth while it is still in the discorger. Barbless hooks are far easier to remove than those that are barbed.

## using forceps

Forceps are the preferred method if the hook is larger than a size 12. They have grooved ends that can grip on the shank of the hook and remove it even if it is deeply embedded in the fish's mouth. They are particularly useful when removing treble hooks from the mouths of predators, because their long, thin arms are able to reach into jaws that are lined with razor-sharp teeth.

## using a keepnet

Keepnets are widely used in match fishing where competitors fish for a set period of time and the winner is the angler who catches the greatest weight. However, many anglers like to keep their fish until the end of the day in order to see what they have caught. Indeed, it can be advisable to use a keepnet when fishing for shoal fish like bream, perch and rudd, because if you release them immediately after capture, they can scare off the rest of the shoal.

Keepnets should be a minimum of 3m (10ft) long with a minimum width of 45cm (18in). When in use, the net should be stretched out so that nearly all of it is submerged in water.

## weighing fish

If you have caught a particularly large fish, you may want to weigh it. Tackle manufacturers make special scales for fish, and you should always use a sling.

Dampen the sling, place the fish in it and then weigh it. Remember to deduct the weight of the sling to get an accurate figure.

---

**unhooking a pike**

Pike are probably the hardest of all coarse fish to unhook because of their tough, bony jaws, lined with rows of tiny but extremely sharp teeth.

1 Remove the fish from the landing net and place it on an unhooking mat.

2 Thread your hand under the gill cover of the fish to get a firm grip. You can use an unhooking glove if you wish.

3 With the other hand remove the treble hooks with forceps. Remember, it is far better to seek advice and watch an expert do this before you attempt to do it yourself.

# essential skills
# legering for **bream**

*Left As bottom feeders, bream are best caught using legering tactics.*

## jargon buster
**Line clip** A small clip on the side of the reel spool that stops the line from coming off the spool once the desired distance has been reached.

Bream are most easily caught by legering. Although they can be taken on a float, their preference for feeding on the bottom, often 27m (30 yd) or so from the bank, makes legering preferable.

## choosing a venue and swim

Bream love lakes, ponds and gravel pits, so this is just the sort of venue to target. Look for a swim with underwater features, such as gravel bars, and plan your attack around these. Gravel bars and other features that aren't visible can be found by casting around with a lead weight (see page 95). Slowly reel it in to find out what the bottom is like. A bumping sensation reveals gravel, which is ideal, whereas difficulty in the retrieve signals weed and silt, which are not.

## choosing your tackle

If you want to leger for bream, choose a quivertip rod (see page 10). This is certainly the most sensitive form of bite indicator and the easiest method for a beginner to grasp. The quivertip indicates exactly when you get a bite, normally by gently pulling round, but will work only if you are positioned correctly. The diagram opposite shows how you should be set up.

Most quivertip rods come with a selection of interchangeable tips, which range in strength and

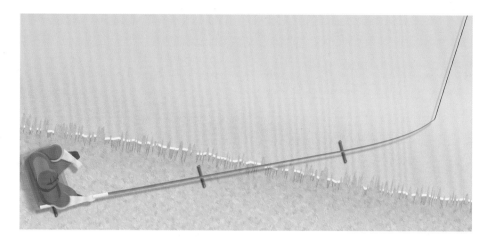

Left *A quivertip provides brilliant bite indication when legering. This is the best way to set the rod up to see those bites.*

are rated in weight divisions. For bream fishing, when bites can be slow and sometimes shy, a soft tip, say 30g (1oz), is usually ideal. A small fixed-spool reel will balance your tackle, and 1.36kg (3lb) main line and a size 16 hook will be good starting points.

## be accurate

Bream love bait – and lots of it. A groundbait feeder is the best way of introducing your feed when quivertipping.

The real key to success is to be careful and accurate – you must hit the same spot cast after cast so that you can build up the swim and keep the bream in the same area. This can be a daunting prospect for the beginner, but a few tricks will help.

All fixed-spool reels have a line clip. Simply cast to your required spot, tuck the line underneath it and reel in. When you cast again, the line will automatically stop at the correct distance.

You can also help ensure your accuracy by picking a marker, such as a tree, on the far bank and casting towards it every time.

## what bait?

Your groundbait feeder will ensure that a regular supply of bait is cast into your swim, and a simple brown breadcrumb should form the base of any mix. Bream love sweet flavours, so try adding a molasses mix to make it even more attractive.

On the hook, bream aren't fussy. Maggots, casters, bread, worm and sweetcorn are all favourites with them.

## key tips

### *for catching bream*

**1** When you've picked your swim, tie on a swimfeeder but no hook. Cast to the same spot with a full swimfeeder five to ten times to build up a bed of bait.

**2** If you haven't had a bite for five minutes, reel in. Bream are shoal fish and move around from area to area. You need a decent amount of groundbait to keep them in the swim.

**3** Try more than one bait on the hook. Sweetcorn and maggot or sweetcorn and worm are great combinations to try.

**4** Don't mess about. Bream are shoal fish and can be caught in numbers, so recast quickly after catching one.

**5** Don't forget those sweet flavourings. Add liquids or powders – pineapple, molasses and strawberry are great – to groundbait.

### swimfeeder rig for bream

**1** open-ended swimfeeder packed with groundbait

**2** 2.27kg (5lb) main line

**3** four-turn water knot for attaching hooklength

**4** 1.36kg (3lb) hooklength

**5** size 18 hook baited with two red maggots

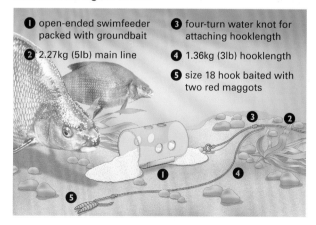

**Above** *A simple yet effective rig like this enables you to get groundbait near your hookbait.*

# bolt-rigging for **carp**

## jargon buster

**Bobbin** A visible bite indicator that sits on the line between reel and the first rod ring.

**Baiting needle** A needle with a small hook on the end, used for threading baits on to the hook of a hair-rig.

The best way to catch a carp is by legering. Because it can take so long to catch a big specimen, floatfishing is often impracticable, and this, coupled with the strong tackle needed to tame this hard-fighting species, makes legering the best choice.

A bolt-rig is a set-up that uses a heavy lead and short hooklength. It works on a simple premise: the fish picks up the bait, feels the weight of the leger and bolts, hooking itself in the process.

## choosing a venue and swim

Carp live in a range of watercourses, but begin on a pond or lake rather than anything bigger because these can present a daunting prospect to the newcomer. Spend time wandering around the fishery, chatting to local anglers and watching the water. Carp are one of the few species that show themselves regularly by crashing on the surface and blowing bubbles, and in warm weather they love to cruise on the surface.

Look for swims with obvious features like islands, overhanging bushes, trees and gravel bars. Don't think you have to cast long distances because carp will often be found in the margins foraging for food.

## choosing your tackle

Above all, your tackle must be strong because carp fight hard. Opt for powerful rods, with a test curve of 1–1.13kg (2lb 4oz–2lb 8oz), and match them with big freespool reels (see page 17). These reels allow line to be taken off the spool without releasing the bale arm, which is essential when bolt-rigging.

Carp move extremely quickly and will often take several metres (yards) of line before the angler has a chance to react. When you get to the rod, simply disengage the freespool facility and lift into the fish – there's no need to strike

**bolt-rig for carp**

❶ 60g (2oz) lead

❷ 5.45kg (12lb) main line

❸ safety clip to allow quick release of the lead if the fish becomes snagged

❹ swivel to attach the hooklength

❺ 4.54kg (10lb) hooklength

❻ size 8 hook with hair-rigged 1.5cm (⅝in) boilie

**Left** *A simple bolt-rig works extremely well for carp.*

## key tips

### for catching carp

**1** If you see a fish crash or roll, cast a bait to it. Lots of fish get caught this way.

**2** Be prepared to put in the hours. Big carp don't give themselves up easily and many get caught while it is dark.

**3** Never ignore the margins. Carp love to patrol the near shelf in search of food.

**4** Put in plenty of bait. Big carp will get through a lot of food in a short space of time.

**5** Hempseed is an excellent – and cheap – way of holding carp in your swim.

because the fish will have already hooked itself. Use a bobbin or an audible bite alarm to indicate the bite. Make sure that the line is strong – anything below 3.63kg (8lb) is unsuitable. Strength is also the first priority when you are choosing a hook, with sizes between 8 and 2 ideal, depending on the size of the bait.

## choosing a bait

Carp will eat a variety of baits, but perhaps the most popular is now the boilie (see pages 46–47). These are best presented on a hair-rig, and although all manner of flavours and colours catch fish, as a rule choose sweeter varieties in summer and fishmeal ones during the winter months. Other classic carp baits include bread, worms, paste, pellets and maggots.

## bolt-rigging for other species

Bream, tench and even roach can be caught in exactly the same way, but rods, reels, line strength and hooks need to be scaled down. The principle behind the set-up remains the same, but fishing with lighter equipment will encourage bites.

# essential skills
# trotting for **roach**

jargon buster

**Pay out line** To feed the main line through your fingers off the reel spool to keep the float running through the water accurately.

**Loafer** A larger version of the stick float. It is designed for bigger baits and more turbulent water.

Trotting is the art of floatfishing on a river, and the word trotting is derived from the action of the float as it 'trots' downriver in the flow. The great advantage of the method is that it presents a bait 'naturally' in the water, as if it were merely a free offering being taken by the current.

## choosing a venue and swim

Not all rivers are suitable for trotting. Sluggish waters of the type normally found in the lower reaches are unsuitable, as are the extremely shallow and fast stretches associated with the upper reaches.

The ideal venue is a medium-paced stretch with a depth of 1–2m (3–6ft). For the method to work, the swim needs to be clear of debris, such as weed and snags, because the float can take a bait right through the middle of the water for anything up to 36m (40 yd) or as far as the angler can still see the tip of the float.

## choosing your tackle

A float rod is essential for trotting. Length is also important, with 3.6m (12ft) the minimum needed, and 4.3–4.6m (14–15ft) the ideal. The extra length

**stick float rig for river roach**

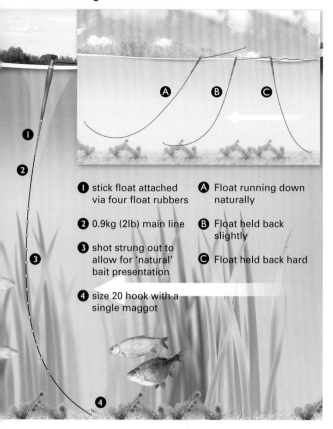

① stick float attached via four float rubbers

② 0.9kg (2lb) main line

③ shot strung out to allow for 'natural' bait presentation

④ size 20 hook with a single maggot

Ⓐ Float running down naturally

Ⓑ Float held back slightly

Ⓒ Float held back hard

**Left** *Trotting is a skill that requires practice, but this rig will help you on your way.*

## key tips

### for catching roach

**1** If the wind is strong, try a technique known as back shotting. Pinch a small shot 15cm (6in) above the float to sink the line and gain better control.

**2** Don't cast too far out when trotting – about 2m (6ft) is the ideal range – because the float won't act naturally and will continually drag itself in an arc towards you.

**3** Use a catapult for feeding. Don't feed by hand unless you're fishing really close in. You want the bait going in a straight line downstream, and a catapult ensures accuracy.

**4** Don't shot all the float's capacity around the base of the float. It should be spread out in a 'shirt-button' style so that the bait acts naturally in the water.

**5** If you want to use bigger baits, choose bigger floats, like loafers. This is a bigger version of the stick float and will give you better control.

makes it easier for you to stop the float from being dragged from its natural course.

Match the rod with a small fixed-spool reel, a closed-face reel or a centrepin (see pages 16–17). Each will work, although more experienced anglers often opt for the centrepin. Roach don't grow big, so the line doesn't need to be more than 1.36kg (3lb), and the addition of a hooklength of say 0.9kg (2lb) will often encourage more bites. As with all species, match the size of the hook to the size of the bait, but most roach anglers wouldn't need anything bigger than a size 12.

Choosing the correct float is vital, and a stick float is ideal. A waggler is attached to the bottom end only, but the stick float is attached both top and bottom with float rubbers. The design enables the float to remain buoyant in fast water and gives the angler better control.

## controlling the float

The angler is trying to achieve an entirely natural presentation of the bait, and the idea is that the hookbait acts as it would do if it were simply being washed downstream with the current. This is achieved by continually paying line once the cast has been made to allow the float to continue its journey. There are a couple of tricks you can use to get more bites.

Try gently pressing your finger on the spool of the reel as the float travels downstream. This will have the effect of holding the float up and ensuring that the bait just trips against the bottom. If you hold back harder still (by pressing your finger firmly on the spool for a second or so), the bait will swing more sharply upwards. Either movement will encourage bites.

## choosing a bait

Roach love a variety of baits, but perhaps the best of them all is maggot. Fish with a single maggot on a size 18 or 20 hook, or, for best results, a double offering on a size 16 hook.

# deadbaiting for pike

## jargon buster

**Drop-off** A part of the swim where the depth changes sharply from shallow to deep water, which is often found in the margins.

**Wire trace** A length of strong, multi-strand wire attached to the lure and main line, designed to withstand the abrasion of the razor-sharp teeth of fish such as pike and zander.

**Pike float** A thick, cigar-shaped float that is large to maintain its buoyancy when a big bait is used.

The pike is Europe's foremost predatory freshwater fish and requires a specialist approach. However, despite their size, pike are delicate creatures and demand great care and attention when caught.

Piking is usually regarded as a winter pursuit, with many waters in Britain allowing anglers to target them only between 1 October and 31 March. In summer, when water temperatures are high and oxygen levels are low, pike exhaust themselves in the fight, so most anglers steer clear of fishing for them at these times. They are, however, a worthwhile target in winter when they often feed at a time when no other species are biting.

## choosing a venue and swim

Pike can be found in a wide range of waters, from lakes, gravel pits and canals to rivers, but it's probably best to settle for a stillwater, where they tend to be more predictable. Look for swims with drop-offs and gravel bars, which is where pike will lurk, waiting to strike at prey. Swims with good marginal cover are also worth concentrating on.

Pike have set feeding times, with dawn and dusk by far the best periods, and be prepared to be mobile, moving swims every couple of hours to find fish.

## choosing your tackle

Pike can grow to more than 18kg (40lb), so your tackle must be strong and robust, while bait must appeal to its basic instinct – eating another fish. Use a rod with a test curve of at least 1.13kg (2lb 8oz) to cast heavy baits. Big freespool reels, capable of holding at least 90m (about 100 yd) of 5.45kg (12lb) line, are also needed.

Probably the best way of fishing for pike is with a float, and specially designed pike floats, which are much bigger and more buoyant than normal varieties of float, are needed to carry the large hookbait.

A wire trace will be essential, because pike have exceptionally sharp teeth, which cut through normal line. Use two sets of treble hooks, normally in sizes between 8 and 4, to which you attach the bait. Remember to take forceps (see page 103).

## choosing a bait

Some waters do not allow the use of live fish as bait, so check the rules before you fish, and remember, just as many pike are caught with dead fish as with live ones.

The choice of bait ranges from small roach, skimmer bream and rudd, to seafish, such as mackerel, smelt and lamprey (see pages 56–7).

**float rig for pike**

Only one other species can be caught with similar tackle and tactics, and that's the zander (see page 79). These are also predatory fish but don't grow so big. Try scaling down the hook size in particular and the bait.

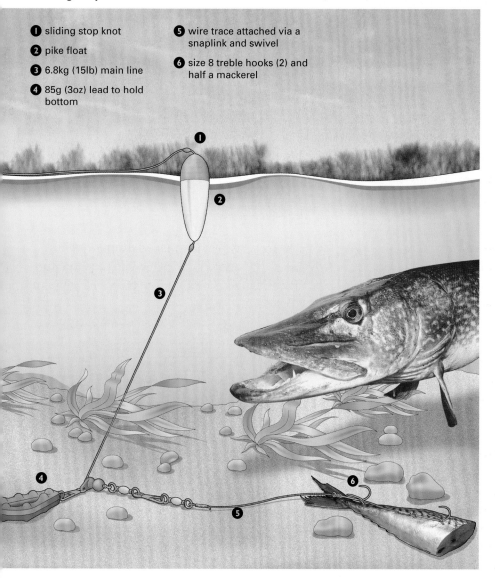

1 sliding stop knot

2 pike float

3 6.8kg (15lb) main line

4 85g (3oz) lead to hold bottom

5 wire trace attached via a snaplink and swivel

6 size 8 treble hooks (2) and half a mackerel

*Left Keep rigs simple for pike. Make sure that they are safe and always use a wire trace.*

## key tips

*for catching pike*

**1** Groundbait for pike but it can work. Simply take a few deadbaits, mash them up into small pieces, bind together with brown breadcrumb and feed in small balls.

**2** Don't be lazy. Travel light and cover as much ground as possible.

**3** Light levels are crucial. Pike feed only when the levels are low, so dawn, dusk or consistently overcast days are best.

**4** Avoid the summer months. Pike exhaust themselves in warmer temperatures and many die after capture.

**5** Pike require a specialist approach with proper rods, reels and, most importantly, specially designed wire traces.

# essential skills
# stalking for **chub**

## jargon buster

**Bait-dropper** A round metal basket with a lid that opens once the dropper hits the riverbed, depositing the feed contained within it.

Chub favour running water, and a river is the best starting point for any beginner, while summer provides the best chance of catching one. When you are stalking, you must be prepared to creep through bankside growth so that you can sneak up on the fish and take it unawares.

## choosing a venue and swim

Clear water is essential for stalking, because you need to be able to see individual fish. This means that you should avoid most big, wide, slow rivers, normally found in the lower reaches, and head for the upper reaches, where the water tends to be shallower and clearer, and the river is narrower.

A classic chub swim will have bankside cover. Chub can be a shy fish, preferring to hide under overhanging bushes and trees before ghosting out on to clear gravel to feed. Look for slightly deeper water. Holes on the riverbed between fast runs are likely holding areas, and the perfect spot will be a combination of the two.

Undercut banks and deep water are also worth investigating too.

## choosing your tackle

When you stalk for chub, you won't normally need bite indication because you will actually see the fish take the bait. You will, however, need a strong rod. Specialist rods, which boast a test curve of about 0.56kg (1lb 4oz), are ideal because they not only have the required power to stop fish burying themselves in weed or underwater snags, but also maintain a soft enough action to be sporting.

A medium-sized fixed-spool reel will balance this type of rod and should be spooled with 2.7kg (6lb) main line. Even if the chub are only about 1.36kg (3lb), you need the extra strength to steer them away from hazards.

Chub have extraordinarily large mouths for their size, so don't be shy when picking a hook.

Again, strength is key, so choose something between a size 10 and 4.

Don't forget your polarizing glasses so that you can see directly into the water without the glare disturbing the view.

## watercraft skills

The best tackle and bait in the world will be useless if you don't approach the river in the correct way. Stealth is vital to success. Approach every likely-looking swim with great caution, walking slowly and keeping off the skyline. Use bankside cover to aid you and be prepared to wait in a swim for a few minutes before a chub gives its location away.

## choosing a bait

The best way to draw chub from their cover is to lay a bed of hempseed a metre (yard) or so from where they are holed up. If the water is low and the area you want to feed is only 3m (10ft) away, this can be done by hand or with a catapult. If it's further out, try using a bait-dropper. As for hookbaits, breadflake, sweetcorn or lobworms are superb chub baits in summer.

*Below* *The perfect chub rig couples simplicity with*
*effectiveness. Keep resistance low with a light link leger.*

## key tips

### *for catching chubb*

**1** The best time to stalk for summer chub is in the evening. As the light levels fall, the fish feel safer and move out from under trees and bushes.

**2** Don't forget your polarizing glasses! If you can't see the chub, you can't catch them.

**3** Wear dark clothing. Stalking relies on stealth, and wearing bright colours will make it more likely that you'll be spotted.

**4** Feed hempseed into several different spots when you arrive. By the time you've fed the last swim, the fish will have had the chance to find the bait in the first one.

**5** Keep rigs simple. Free-lining – just offering the hookbait with no lead on the line – is often the best approach of all.

## other species

Barbel are another river species that respond well to stalking. Be aware that they grow bigger and fight harder than chub, so increase the line strength accordingly. You will also need a heavier rod, say one with a 0.68kg (1lb 8oz) test curve.

**leger rig for chub**

❶ three SSG shot to hold bottom

❷ 2.7kg (6lb) main line

❸ four-turn water knot to attach leger link

❹ size 10 hook baited with breadflake

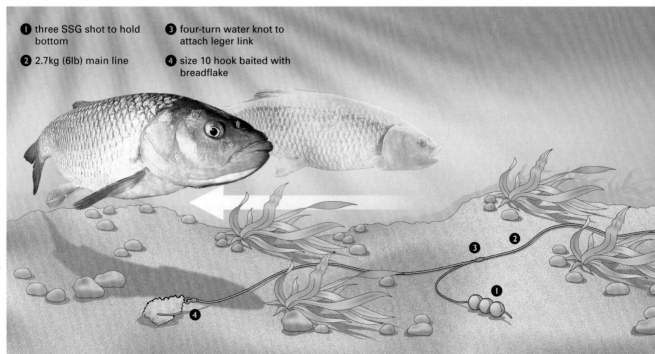

# essential skills
# floodwater fishing for **barbel**

Barbel are probably easiest to catch in winter, but only at certain times. They go on feeding binges when the water levels rise, and their normal caution is apparently forgotten. The best conditions come when the temperature rises after a cold spell, causing rain to fall and river levels to rise. The rain muddies the water, and barbel start to feed.

## jargon buster

**Flooded river** This is when a river is carrying more water than usual, normally after heavy rainfall. Only fish if it is safe to do so and the bankside is stable.

**Hold bottom** Leads need to be heavy enough to remain stable and keep the bait in place. They need to be able to 'hold bottom'.

## choosing a venue and swim

A river should be your venue, but picking a swim isn't quite so easy. Finding a swirling, turbulent, chocolate-coloured mass of water that has replaced a normally sedate stretch of river is not easy, but it can provide some excellent sport.

The first thing is to make a mental note of the river before flooding. It's vital that your bait is on the original riverbed and not on part of the flooded bank, so sketch the contours in summer when levels are low, using permanent landmarks like trees or bankside structures. Once you've pinpointed a particular area, choose a swim where the water is smooth, which indicates a flat riverbed. The behaviour of the water on the surface reflects the underwater contours – boiling eddies reveal an uneven, rocky bottom, which is useless for barbel.

## choosing your tackle

Flooded rivers are almost as powerful as barbel themselves, so your tackle must be strong and robust. Choose a rod with a test curve of at least 0.9kg (2lb) and match this with a freespool reel that should be loaded with at least 3.63kg (8lb) main line. Bite indication is largely unnecessary – barbel hit the bait hard and will run immediately on feeling the weight of the lead.

**floodwater rig for barbel**

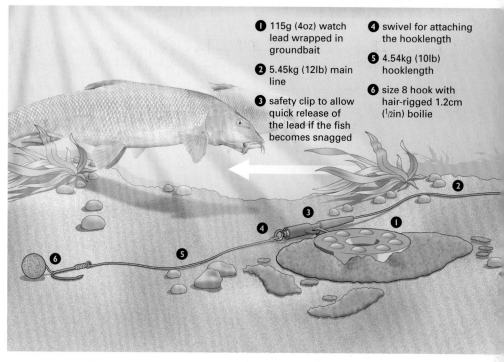

❶ 115g (4oz) watch lead wrapped in groundbait

❷ 5.45kg (12lb) main line

❸ safety clip to allow quick release of the lead if the fish becomes snagged

❹ swivel for attaching the hooklength

❺ 4.54kg (10lb) hooklength

❻ size 8 hook with hair-rigged 1.2cm (½in) boilie

## key tips

### for catching barbel

**1** The best time to floodwater fish is after several days of heavy rain in the middle of a cold spell. Wait for the levels to begin to drop before fishing.

**2** Smelly baits are the best to use in flooded conditions.

**3** Make sure that you use plenty of weight to hold bottom. It's essential that the bait stays static.

**4** Loose feeding is a waste of time in flooded water. The bait will never hit the bottom before it is washed away.

**5** Take care with your catch. Barbel are delicate and require careful handling when they are released.

## target eels

**Apart from barbel, the only other fish to feed heavily in flooded conditions are eels. Try using worm as the hookbait if they are your target.**

In fast water you must keep the rod pointed high in the air so that as much line as possible is kept off the water, otherwise the force of the flow will continually move the lead.

A heavy lead is essential to hold bottom, and at the height of a flood, anything between 85 and 170g (3–6oz) could be needed. Hooks must be between a size 8 and 4 depending on the type of bait you choose.

## choosing a bait

The key to floodwater barbel baits is smell. In coloured water fish won't be able to see the bait, so it's vital that your hookbait is both big and potent. Luncheon meat is a favourite barbel bait, and anglers will use chunks as large as matchboxes. The meat has a high oil content, which will waft down the current and attract the fish.

Boilies are a good alternative; choose fishmeal flavour in winter. To make the bait more attractive, dip the boilies in one of the flavours you'll find in tackle shops – crab, shrimp and mussel are some of the best to use in winter.

## how to release barbel

Despite their power, barbel are fragile fish, and you must exercise special care and attention when you release them. This is especially true in summer when both water and oxygen levels are low, but even in winter anglers should follow a few simple guidelines to ensure their safe return.

Immediately after capture, it's best to leave the fish in the landing net for a few minutes to recover. Never take it out immediately, unhook it and release it straight away. When it's had time to regain its strength, you can let it go. Always point the barbel with its head into the current because the flow will aid recovery, and only let go when the fish struggles to break free.

# essential skills
# lure fishing for perch

## jargon buster

**Sink and draw** A term used to describe how to retrieve a lure. Simply cast, allow the lure to sink, then draw it back with a couple of short turns on the reel. Repeat and recast.

Perch can be caught on traditional baits, like maggots, worms and fish, but perhaps the most exciting way is with a lure. They are predatory fish, hunting their prey by sight, and will attack a lure that is worked through the water to mimic a distressed fish.

## choosing a venue and swim

Lure fishing can be done on virtually any water. All canals, lakes, gravel pits and rivers hold perch, but for big fish – those weighing more than 1.36kg (3lb) – perhaps big lakes and pits are best, although they do appear on rivers too.

Perch love cover in the form of overhanging bushes, snags, bridges and any other structures in the water, and it is where they wait in ambush for passing fish.

Be prepared to move around because there's little point in staying in a swim for any longer than five minutes. Perch feed on instinct, and if they don't take the spinner straight away, it's unlikely that they will at all.

Don't try to cast your lure too close to snags until you have more experience. Lures are expensive and you don't want to loose them.

## choosing your tackle

Purpose-built lure rods can be bought from tackle shops. They are shorter than float or leger rods because you do not need the extra length to control a float or launch a bait 90m (100 yd). They should be coupled with a fixed-spool reel. Line strength of 2.7kg (6lb) will be capable of landing any fish that you're likely to encounter, as well as being durable enough to deal with the constant recasting of a lure.

Don't forget a wire trace (see page 111). Perch haven't got teeth to cut through the line, but pike have, and they may well pick up the lure even if it isn't intended for them.

## choosing a lure

There are so many lures to choose from that picking the right one can be difficult. The first

**lure fishing for perch**

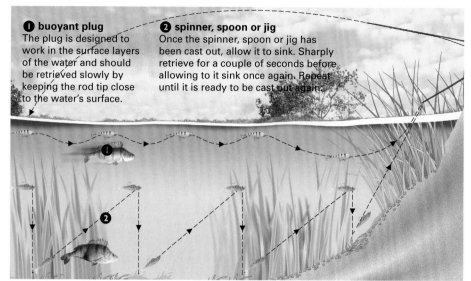

**❶ buoyant plug**
The plug is designed to work in the surface layers of the water and should be retrieved slowly by keeping the rod tip close to the water's surface.

**❷ spinner, spoon or jig**
Once the spinner, spoon or jig has been cast out, allow it to sink. Sharply retrieve for a couple of seconds before allowing to it sink once again. Repeat until it is ready to be cast out again.

*Left* *The key to success is working the lure correctly. And remember, pick the right one for the job.*

thing to consider is the size. Many lures are designed for pike and are much too big for perch, so choose something smaller. The lures available to use for perch are summarized below, but see also page 57.

## Spinner

So called because part of the lure revolves, this is probably the most popular lure for perch fishing; usually coloured silver or gold, it carries a single treble hook.

## Spoon

Slightly longer and more colourful than a spinner, this has a single blade attached to a treble hook; it is often shaped to resemble a fish.

## Plug

Different from spinners and spoons, it can be worked at different levels of the water; some float, some work at mid-depth and some dive deep. They are more bulbous than spinners or spoons and come in a variety of colours.

## Jig

A relatively new introduction to Europe from the USA, this has a weighted rubber body so that it sinks. It has a single hook and gets its name from its action underwater. It requires the angler to continually flick the tip of the rod.

## key tips

### for catching perch

**1** Cover lots of ground. Being mobile will mean you cover more water and catch more fish.

**2** Don't forget the wire trace. Perch don't have teeth, but pike do, so go well prepared.

**3** Choose the right lure. Most are made for pike, and these are too big for catching perch.

**4** Take care when you are casting. You want to be close to snags, but not too close. Lures are expensive and you don't want to lose them.

**5** Try lures that contain an internal rattle. They vibrate underwater, sending out a noise that alerts fish.

## target pike

Pike also love lures. If you are targeting them, increase the line strength, use a wire trace and increase the size of lure. If you are after pike, don't forget the forceps!

## how to work a lure

The best approach for a beginner is the 'sink and draw' method. This involves casting, allowing the lure to sink for a few seconds, retrieving it with a couple of sharp turns on the reel, then allowing it to sink again. Repeat the procedure until you reach the margins, but don't be in a hurry to remove it – perch follow the lure to the water's edge before taking. You are trying to mimic the behaviour of a dying fish, so a straightforward cast-and-retrieve style is unlikely to work.

# essential skills
# using the lift method for **tench**

All you need for this classic method of catching tench in the summer months are a rod, reel, float, hook and some bait. Although bigger tench tend to be caught by those who choose to leger at range, there remains something special about catching one of Europe's most sought-after species on the float.

## choosing a venue and swim

Tench are traditionally found in stillwaters, and the best choice of venue is a pond, lake or small gravel pit.

Tench love to patrol the margins, and the ideal swim will be one where the margin gently shelves to a depth of about 2m (6ft) to a weed-free gravel bottom. This is exactly the kind of place where tench will forage for food at the base of the shelf. If the swim has bankside cover in the form of an overhanging tree or bush, so much the better.

Using a rake can give you an edge. Tie a rake head to the end of a piece of rope and throw it into the swim. Retrieve and repeat several times until the head comes back clean. Not only does this stir up the bottom and release lots of natural food into the water, but the noise will attract the inquisitive tench into the area.

## choosing your tackle

Tench are hard-fighting species, so opt for a 0.56kg (1lb 4oz) test curve rod. Casting a great distance is unnecessary when using the lift method, so power is the first priority. Choose a medium fixed-spool reel to balance the rod and load it with 2.7kg (6lb) line. Any tench over 0.9kg (2lb) will fight hard, so make sure that the hook you use is a strong one. A size 14 or 12 hook will suit smaller baits, while a size 8 or 10 will be better for bigger ones.

## lift method for tench

1 make sure that the rod tip is set well back to avoid scaring the fish

2 attach a straight waggler top and bottom with float rubbers; the float will then lift when the tench picks up the bait

3 3.63kg (8lb) main line

4 single large shot, either AAA or SSG

5 size 8 hook with a single lobworm

*Left The lift method is simple but deadly. The float will rise, not disappear.*

## how does the lift method work?

The method ensures that the bait remains static on the bottom, but also employs the float to indicate a bite, which increases its sensitivity. Simply fix a straight waggler-style float with two float rubbers at the top and bottom. Plumb the depth and place a heavy shot, say two AAAs or one SSG (see page 21), on the lake bed, where it cocks the float. Then tie on the hook 15–20cm (6–8in) away.

When the tench picks up the bait, it dislodges the shot and lifts the float out of the water, hence the name of the tactic. Have a look at the diagram above to see how to setup the perfect lift method.

## choosing a bait

Tench aren't particularly fussy about what they eat, but there are several summer favourites. Sweetcorn, breadflake and lobworms are all excellent baits, with maggots and casters good alternatives.

## target carp

Carp love to cruise along the margins and respond to the lift method. However, they are bigger, more powerful fish, so you'll need to use a heavier rod, bigger reel, thicker line and a stronger hook.

## key tips

### for catching tench

1 Don't forget to use a rake. Tench love to root around for food, and if you stir up the bottom you'll encourage them into your swim.

2 Hempseed is an almost universal attractant to all coarse fish, and tench are no exception. Feed hemp and fish your chosen hookbait over the top.

3 Take a walk around your venue at dawn or dusk. Tench give themselves away by rolling and blowing feeding bubbles to the surface, and knowing where they are will give you a head start.

4 Polarizing glasses will enable you to see right into the margins of the lake. Look out for holes in the weed or for clear gravel areas.

5 Make sure that you plumb the depth accurately. The weight must be at absolute dead depth for the method to work well.

# 50 best fishing tips

Now that you've learned about tackle, how to set it up, what baits to use and where fish live, you're ready to go fishing. With those basics under your belt, you will certainly catch a range of species, but there is so much to learn in angling that you could spend a lifetime picking up new ideas. Here are 50 of the best fishing tips that will give that extra edge when you're out on the bank.

## legering

• Make sure that you are using the correct type of quivertip to suit your style of fishing. For example, a soft, light tip made from fibreglass is ideal for shy-biting roach and skimmers, while a stiff, carbonfibre tip will cope with powerful, fast-flowing rivers.

• When faced with fast, boiling river swims where a normal leger weight will roll away in the current, try flattening the lead with a hammer. This will allow the lead to sit flush on the bottom and not be dislodged by the flow.

• Use the line clip on your reel when you use a swimfeeder. This will allow you to hit the same spot when casting on lakes and to the far banks of rivers, thus concentrating your feed in one spot.

• River flows and tows on lakes can alter over the course of a session, so it's worth carrying some extra add-on leads with you. Clip these on the feeder one at a time until the balance of your rig is just right once again.

• In clear water it is well worth camouflaging your feeders and leads so that they don't stand out to fish. A waterproof permanent marker or waterproof paint will do the job perfectly and can be removed with a knife or a pair of scissors afterwards if necessary.

## floatfishing

• When you're fishing a waggler with corn or meat at distance, strike your bait off before reeling in. This is a superb way of getting bait in your swim when it is out of the range of the catapult.

**Above** *Be sure to shot your float correctly – too high and you won't hit many bites.*

• A plummet is as vital as the bait. Take your time to find the features in your swim.

• Shot your float correctly. Only the tip should be showing above water so that even the shyest bite will result both in its disappearing completely and in a hooked fish. Less float sticking out also means there is less chance that it will be affected by the wind.

• Use float adaptors. That way you can quickly change float size should conditions such as the wind or flow change without having to break down and tackle up again, which can cost you valuable fishing time.

• Make sure that the tip of your float is visible against the reflections on the water. For example, if you're fishing in light-coloured water, use a black tip; if the reflection on the water is dark, a red or orange-coloured tip will be better.

**Above** *Be accurate with your feeding, otherwise it will do more harm than good.*

## feeding

• In summer try and drive the fish into a feeding frenzy by adding bait little and often. The best weights aren't always taken on the greatest amount of bait.

• Pre-baiting a swim with pieces of fish every other night over a period of a week or so can be a successful tactic.

• When the fishing is hard, try casting just beyond your baited area. Often wary fish will hang back, so it's worth having an extra cast to see.

• Feed more than one swim to give you more choice. Why not feed close in for floatfishing and further out with a swimfeeder? If one is not producing you always have the other to rely on.

• Be patient when you see feeding fish. When you've put down a bed of hempseed and barbel and chub appear over it, wait for the fish to gain confidence before casting your hookbait. If you throw the bait in too early, you'll spook any fish in your swim.

## bait

• Bread is a superb bait for big roach. Try using it during the first and last hours of daylight, which is when the fish have more confidence to grab a big offering.

• Always riddle groundbait mixes. This distributes the wettest lumps evenly throughout the mix and makes certain that the fish don't fill up to soon.

• Using flavoured baits will give you an edge. A fruity flavour on maggots works well in summer, whereas a spicy one is better for winter. Remember to rub a little on your hands to take away the 'human' smell.

• Never leave home without hempseed. Whether it's used as hookbait or as loose feed, there is no finer bait for attracting roach.

• Dead maggots can be the answer when you are fishing in weedy water. Live ones tend to crawl away quickly, leaving no feed around your hookbait. Moreover, dead maggots, when frozen, can be easily flavoured.

## watercraft

• If you've only got a few hours to go fishing, don't waste them. In summer, go after school or work, because those evening sessions can be the most productive as the fish become more active.

• Watch the water before setting up. Small fish breaking the surface is a sure sign of feeding pike, and grebes diving and feeding in the same area shows that they have found a shoal of food fish.

• Fish tight to overhanging cover, such as reeds and bushes, where fish seek a safe haven from aerial predators

**Above** *Dawn or dusk are great times to be on the bank because it's when fish feed.*

• Even on huge pits never ignore the margins. Species like pike and zander can often be caught quite close to the bank, particularly as the light begins to fade.

• Always keep a close eye on the atmospheric conditions, especially the pressure. In high pressure conditions the fish feed for only short periods, but in low pressure they will feed for longer, so you can be more attacking in your fishing methods.

## roach

• Big roach react to light. Try targeting specific parts of the day to maximize your chances of success. The first hours of daylight and then again at dusk are the best times.

• Winter is a better time of year if you want to catch specimen roach. The colder weather means that small baits aren't as likely to be attacked by small fish, and the roach themselves are at their optimum weight in preparation for spawning.

• Feeding one type of bait and fishing another on the hook is important. Examples would be feeding casters and fishing maggot, or feeding squatt and hooking pinkie. The fish tend to pick out the change of bait among the loose feed.

• The perfect roach conditions on a river are when visibility into the water is 23–30cm (9–12in), the water temperature is 5°C (41°F) and rising, and there are low light levels on a mild day.

**Above** *Pike can be difficult to unhook. Forceps are essential – seek advice before doing it yourself.*

• When you're fishing the bomb for roach, always cast around the area you are feeding. Spray maggots in a diameter of about 2m (6ft) and cast to different spots in that area. This tends not to spook the roach.

## bream

• A common trick to try when you're using a swimfeeder when fishing for bream is to swap to a leger bomb when bites dry up. Fish can become wary of a swimfeeder landing near them time and time again, but a change to the lead will dispel these inhibitions and get you catching again.

• Pre-baiting is essential if you are aiming for a massive catch of bream. Introduce groundbait, corn, pellet and mashed bread for a few evenings and then arrive the following morning. The fish should be queuing up to be caught.

• Bream are shoal fish, so if you catch one, you'll probably get another. Therefore don't waste time and recast as quickly as you can.

• Bream love pellets. Most anglers use them for carp and barbel, but they are great for keeping bream in your swim. Add small 3mm (⅛in) pellets to groundbait.

**Above** *It might be cold, but winter is the best time to catch a big roach.*

• Mini-boilies are a great alternative for bream. Fishmeal-flavoured ones tend to work the best, so try two on a hair-rig.

## pike

• Pick the right lure for the conditions. As a rule, surface, shallow-diving and buoyant lures are ideal in warm water, while sinking and deep-diving lures are better when it's cold.

• When you are lure fishing, don't snatch the lure from the water at the end of the retrieve. A pike may have followed it, so just swish it under the rod tip to induce a last-gasp take.

• Pike might be fearsome-looking creatures, but they are extremely delicate. Be responsible and unhook, weigh and photograph the fish as quickly as possible.

• Whether you're lure fishing, deadbaiting or livebaiting, you must always use a wire trace. Pike and zander have extremely sharp teeth, which will easily cut through monofilament.

• Striking twice ensures a good hook hold when you are fishing for pike or zander, especially at long range. Strike first when you get the bite, tighten up and strike again to make sure that the hooks are firmly set.

## barbel

• Make sure that you nurse barbel back to full health before release. Give them a breather in the net or carefully hold the fish's tail until the fish gives a violent flap and struggles to get away.

• If barbel and chub are the target, travel light. A rod, landing net and bait bucket are all that's needed for a session. The more ground you cover, the better your chances of catching fish.

• Weirs contain superb fish-holding features. Under the sill and in the whitewater you'll find barbel, whereas in the slacker water at each side there will be species like chub, bream and pike.

• Don't be put off by a flooded river – it can provide excellent sport. Barbel in particular feed heavily when the river's up and coloured, but look for quiet slacks out of the main torrent.

• When the river is in flood, use the biggest, smelliest bait you can find. Luncheon meat is

brilliant because it releases oils, but spicy boilies also work well.

## carp

• Be prepared to put in the hours. Carp are often at their most active as dusk falls, through the night and then at first light. You won't catch them if you're in bed.

• Make sure that you use enough weight to make the rig effective when you are bolt-rigging for carp. Try lifting the rig with a finger under the hook point. If it hurts, you've got it right!

• Remember that carp are the hardest-fighting of all coarse fish, so you need to tackle up accordingly. Don't consider fishing without at least 5.45kg (12lb) main line, and 6.8kg (15lb) will be safer. You're not fishing to lose fish.

• Don't think that the only bait that will catch carp is a boilie. While they have proved to be excellent carp-catchers, there are plenty of alternative baits that may produce a better response. Worms, bread, sweetcorn and meat are all worth a try.

• When you are fishing for river carp, look for features that provide both shelter and food. Try baiting up bridges, water outlets, near waterlilies, under overhanging trees and near moored boats.

**Above** *Carp fishing can be tough, so make sure that your equipment is up to the job.*

# fishing in Europe

Different countries have different rules and regulations about who can fish and when and where. Always check before you fish – local tackle shops will be a good starting point.

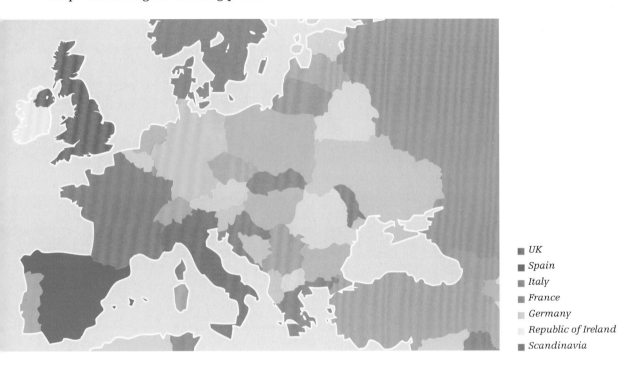

■ UK
■ Spain
■ Italy
■ France
■ Germany
■ Republic of Ireland
■ Scandinavia

## France

### is there a season?
There is no closed season on any venue, so you can fish all year round.

### do I need a licence?
There is no rod licence as such, although each region, or *département*, issues different permits relevant to the fisheries within their remit. Some privately owned lakes do not need any permits, so check before you fish. Permits are available from local tackle shops, bars and cafés.

### what are the rules?
Rules are few and far between for fishing on general rivers and lakes, but many private lakes enforce rules strictly, especially for their stocks of specimen carp. Relevant fishery rules will be available on request when you buy your permit, so don't be afraid to ask.

## Germany

### is there a season?
While there is no closed season for species such as bream and roach, there is one for predators, like pike and zander. The predator closed season runs from April to June for zander, and from January to April for pike.

### do I need a licence?
Germans must sit an exam before they can gain a fishing licence. This is the Certificate of Competence, and anglers must prove their knowledge of, among other things, fish, techniques and the environment.

### what are the rules?
These vary from region to region. In addition, in some parts of Germany no fish can be returned alive, so find out exactly what the rules are before casting out.

## Italy

### is there a season?

There is no closed season, but each species has a prohibition period around the time of spawning (usually from April to June), and during this period you are not allowed to fish for these species. Local tackle shops will advise you because the periods vary from region to region.

### do I need a licence?

Italian fishing permits last for six years. Visit the local licensing agency with passport photos and proof of ID, and complete a form that must be processed at the local post office.

### what are the rules?

You can remove fish from venues in some regions of Italy, but strict size limits are imposed on different species, so make sure that you check before doing so. Night fishing is rare but not illegal. Check in the local tackle shop.

## Republic of Ireland

### is there a season?

There is no closed season, but there are some restrictions for game fishing. Check in local tackle shops for details.

### do I need a licence?

You do not need a licence or permit of any sort, and fishing is free.

### what are the rules?

The rules in the Republic are similar to those of the UK in terms of removal of fish. Livebaiting is banned, but night fishing, although not common, is allowed in most places.

## Scandinavia

### is there a season?

Fishing is permitted all year round, but in winter entire lakes and rivers can freeze and ice fishing becomes popular.

### do I need a licence?

Anglers will need a local licence to fish in both Denmark and Sweden, and these cost around 16 Euros from tourist offices. Most towns and cities have an office, but tackle shops will also be able to supply licences.

### what are the rules?

The rules are fairly relaxed because many native anglers catch fish for the table and removing fish is, therefore, quite normal. Salmon and trout fisheries may have different rules, so check when you buy your permit. Some lakes in Denmark restrict the amount of groundbait allowed.

## Spain

### is there a season?

Generally there is an open season. However, some lakes impose a fishing ban in particularly hot weather when oxygen levels in the water drop to dangerously low levels.

### do I need a licence?

Rod licences are required; these licences can be purchased from local tackle shops or local authorities in each province.

### what are the rules?

Because fishing is free in Spain, there aren't too many rules. However, on the Ebro and many other rivers there is a night fishing ban, and lakes often enforce their own rules.

## United Kingdom

### is there a season?

Closed on rivers between 14 March and 16 June, but open all year round on most canals and stillwaters. Check before fishing.

### do I need a licence?

Yes. A rod licence is essential for all those over the age of 12. They can be purchased from the Environment Agency from post offices, as well as via the internet.

### what are the rules?

It is generally accepted that all fish caught in the UK should be returned to the water. Night fishing is not normally allowed unless permission is granted beforehand.

# index

**a**

action, rods 10, 11
additives 38, 39, 45, 47
alarms, bite 24
anti-reverse levers 14, 15, 16, 100
Avon-style rods 101

**b**

backing, reels 28
backwinding 100
bait 42-57
   bait boxes 22
   boilies 46-7
   bread 44-5
   catapults 23
   deadbaits and lures 56-7
   feeding 121
   maggots and casters 48-9
   matching to hook size 49
   mixing groundbait 38-9
   pellets, paste and luncheon meat 54-5
   seeds, pulses and nuts 52-3
   swimfeeders 21
   worms and other natural baits 50-1
   see also individual fish
baiting needles 106
bale arms, reels 14, 15, 16
banks, undercut 69, 99
banksticks 24
barbed hooks 20
barbel 60-1, 123
   floodwater fishing 114-15
   releasing 115
   in rivers 91
   stalking 113
   in weir pools 93
barbless hooks 20
bedding in, line 28
bends, choosing a swim 99
bite 30, 31
bite alarms 24
blank, rods 11
bleak 82
block-end feeders 35
blood knot, half-tucked 41
bloodworms 49, 51
bobbins 25, 106
bodied wagglers 33
boilies 46-7
   hair-rigs 47
bolt-rigging for carp 106-7
bombs 21, 34, 35
boxes: bait 22
   seat 23
   tackle 22

brandlings 50
Brasem 38, 39
bread 44-5
bread paste 55
breadcrust 44
breadflake 44, 45
breadpunch 44
breaking strain (BS), lines 20, 21
bream 62-3, 122-3
   in canals 89
   deadbait 57
   fixed rigs 35
   in gravel pits 95
   in lakes 87
   legering for 104-5
   in rivers 91
   shotting floats 32
   in weir pools 93
bridges, choosing a swim 99
butt section, rods 11

**c**

canals 88-9
carbonfibre 10, 11, 18
carp 7, 123
   bolt-rigging 106-7
   in canals 89
   common carp 66-7
   crucian carp 71
   grass carp 83
   in gravel pits 95
   in lakes 87
   lift method 119
   in weir pools 93
carp poles 18-19
carp rods 13, 101
casters 48, 49
casting 30-1
   centrepin reels 17
   leger weights 34
catapults 23
catfish: in gravel pits 95
   wels catfish 64
centrepin reels 17
chairs 24
cheesepaste 55
chopped worm 51
chub 68-9
   in rivers 91
   stalking 112-13
   in weir pools 93
closed-face reels 16
clutch, playing a fish 100
cockles 51

**d**

dace 65, 68, 91
deadbaits 56-7

catching pike 110-11
deep slacks 99
depth, plumbing 98
discorgers 23, 102-3
dotting down 32
drag, reels 14, 15, 16
drop-off, swim 110

**e**

earthworms 50
eels 78
elastic, pole fishing 18, 19, 36
elderberries 51
equipment: poles 18-19
   reels 14-17
   rods 10-13
   tackle sundries 20-5
Europe, fishing in 124-5

**f**

feeders 21, 121
   block-end feeders 35
   groundbait feeders 35
   method feeders 35
   swimfeeders 34
fibreglass 10, 11, 18
fish: finding 85-95
   landing 100, 101
   playing 37, 100-1
   types of 60-83
   unhooking 23, 102-3
   weighing 103
fishmeal pellets 54-5
fixed rigs 35
fixed-spool reels 14-15, 16
flavoured bread 45
flicktips 18, 19
float rods 12
floatfishing 120
floats 10, 21
   deadbaiting for pike 110
   lift method for tench 119
   pole floats 37
   shotting 32, 33
   trotting 14, 109
   types of 33
floodwater fishing, barbel 114-15
forceps 25, 103
France, fishing in 124
freespool reels 14, 17
freezer baits 46

**g**

game fishing 6, 17
Germany, fishing in 124
glasses, Polaroid 25, 113
grass carp 83
gravel bars 99, 104

gravel pits 94-5
gravel runs 99
grayling 83, 91
groundbait: additives 38, 39
  feeders 35
  mixing 38-9
gudgeon 82

h
hair-rigs, boilies 47
half-tucked blood knot 41
handles: reels 15
  rods 11
hempseed 52, 53
hold bottom 114
hooklengths 20, 21, 40
hooks 20
  anatomy 20
  deadbaiting for pike 111
  floodwater fishing for barbel 115
  hooking maggots and casters 49
  hooking worms 51
  lift method for tench 118
  matching bait to 49
  mounting breadflake 45
  playing a fish 101
  spade-end hooks 41
  stalking chub 112-13
  striking 31
  unhooking fish 23, 102-3
  using luncheon meat 55
  using pastes 55
hybrid fish 77

i
ide 82
insert wagglers 33
Ireland, fishing in 125
islands, choosing a swim 99
Italy, fishing in 125

j
jigs 57, 117

k
keepnets 22, 103
knots 40-1

lakes 86-7
lamprey, as deadbait 57
landing a fish 100, 101
landing nets 22
lead: floodwater fishing 115
  plummets 23, 32
  shot 21
leger rods 12
leger weights 10, 21, 34
legering 10, 120

bream 104-5
  setting up 34-5
  stalking chub 113
licences 124-5
lift method, tench 118-19
line capacity, reels 28
line clips 14, 16, 104, 105
lines 21
  bedding in 28
  bolt-rigging for carp 107
  breaking strain 20, 21
  casting 30-1
  centrepin reels 17
  hooklengths 20, 21, 40
  monofilament 20, 21
  paying out 108
  playing a fish 100, 101
  spooling reels 28-9
  tangling 15
  trotting for roach 109
link legers 35
loafers 108
lobworms 50, 51
luggage 23
luncheon meat 55
lure fishing, perch 116-17
lure rods 116
lures 57

m
mackerel, as deadbait 57
maggots 38, 48-9
maize 53
margin poles 19
margins 18
mash 45
match anglers 6
match rods 101
mats, unhooking 102
method feeders 35
minnows 57, 83
monofilament 20, 21

n
nase 83
natural baits 50-1
needles, baiting 106
nets: keepnets 22, 103
  landing a fish 100, 101
  landing nets 22
nuts, as bait 52-3

'on the drop' 32
overhand loop knot 40

p
particle baits 52-3
pastes 55

paternoster rigs 34, 35
paying out lines 108
peanuts 53
pellets 54
perch 72-3
  in canals 89
  in gravel pits 95
  in lakes 87
  lure fishing 116-17
  in rivers 91
  in weir pools 93
pike 7, 74-5, 123
  in canals 89
  deadbaiting 110-11
  in gravel pits 95
  in lakes 87
  lure fishing 117
  in rivers 91
  unhooking 103
  in weir pools 93
pike floats 110
pike rods 13, 101
pikeperch 79
pinkies 49
playing fish 37, 100-1
plugs 57, 117
plumbing the depth 98
plummets 23, 32
pods, rod 24
Polaroid glasses 25, 113
pole fishing 36-7
pole floats 37
pole rollers 37
poles 18-19
ponds 86-7
pools, weir 92-3, 99
prawns 51
pulses, as bait 52-3
pumping the fish 101

q
quivertip rods 10, 12, 101, 104-5

r
redworms 51
reedbeds 99
reel seat, rods 11
reels 14-17
  anatomy 15
  backing 28
  bolt-rigging for carp 106
  deadbaiting for pike 110
  line capacity 28
  playing a fish 100, 101
  spooling 28-9
  tangling 15
  types of 16-17
riddles 38

rigs: bolt-rigging for carp 106-7
  deadbaiting for pike 111
  fixed rigs 35
  floodwater fishing for barbel 115
  leger rigs 34-5, 105
  lure fishing for perch 117
  pole fishing 37
  running rigs 35
  stalking chub 113
  trotting for roach 109
rings, rods 11
rivers: choosing a swim 99
  finding fish 90-1
  flooded 114-15
roach 76-7, 122
  in canals 89
  as deadbait 57
  in gravel pits 95
  in lakes 87
  in rivers 91
  trotting 108-9
  in weir pools 93
roach poles 18, 19
rod pods 24
rod rests 24
rods 10-13
  anatomy 11
  bolt-rigging for carp 106
  casting 30-1
  deadbaiting for pike 110
  floodwater fishing for barbel 114-15
  legering for bream 104-5
  lift method for tench 118
  lure fishing for perch 116
  playing a fish 100, 101
  stalking chub 112
  test curve 10
  trotting for roach 108-9
  types of 12-13
rollers, pole 37
rudd 70, 77
  in canals 89
  in gravel pits 95
  in lakes 87
running rigs 35

**s**
scales, handling fish 102
scales, weighing fish 25, 103
Scandinavia, fishing in 125
Scopex 38, 39
seafish 6
seat boxes 23
seeds, as bait 52-3
shelf-life boilies 46
shipping out poles 36
shot 20, 21, 35
shotting floats 32, 33
sidestrain, playing a fish 100-1
silver bream 63
sink and draw, lures 116
skimmer bream 57, 62
slabs, bream 62
sliding stop knot 41
slime, handling fish 102
slugs 51
smelt, as deadbait 57
snags 18, 99
snaplinks 34
spade-end hooks, tying 41
Spain, fishing in 125
specialist rods 13
spinners 57, 117
spinning 10
spinning rods 13
spooling reels 28-9
spools 14, 15
spoons 57, 117
squatts 49
stalking chub 112-13
stick floats 33
stillwater, choosing a swim 99
straight crystal wagglers 33
striking 30, 31, 36
sweetcorn 52, 53
swim, choosing 98-9

swimfeeders 10, 21, 34, 105
swivels 34

**t**
tackle boxes 22
tackle sundries 20-5
tares 53
tench 80-1
  in canals 89
  in gravel pits 95
  in lakes 87
  lift method 118-19
  in weir pools 93
test curve, rods 10
tiger nuts 53
treble hooks 20
trees, overhanging 99
trotting 14, 108-9

**u**
umbrellas 25
undercut banks 69, 99
unhooking fish 23, 102-3
United Kingdom, fishing in 125
unshipping poles 36

**w**
wagglers 33
watch leads 34, 35
water, plumbing the depth 98
water knot 41
watercraft 121-2
weed, clear patches in 99
weighing fish 25, 103
weights, leger 10, 21, 34
weir pools 92-3, 99
wels catfish 64
whips 19
wire traces 110, 111, 116
worms 50-1

**z**
zander 79

# acknowledgements

With grateful thanks to **Drennan International, Dynamite Baits, Svendsen Sport, Martin Bowler, Terry Lampard and Duncan Charman.**

**All Photography: Angling Times/Emap.**
Except for the following:
**Octopus Publishing Group Limited and Angling Times**/Steve Partner 12 top, 12 bottom, 13 top left, 13 bottom, 15, 16 top left, 16 top right 17 top left, 17 top right, 19 top left, 19 top right, 19 top centre right, 19 top centre left, 21 top left, 21 top right, 21 bottom right, 21 bottom centre, 21 top centre, 22 top left, 22 top right, 22 bottom right, 23 top, 23 bottom left, 24 top left, 24 top right, 24 bottom right, 24 bottom left, 25 top left, 25 top right, 25 bottom right, 29 top left, 29 top right, 29 bottom right, 29 bottom left, 33 centre left, 33 centre right, 33 bottom right, 33 bottom left, 35 bottom left, 37 centre left, 37 bottom left, 46 bottom left, 47 top left, 50 bottom right, 52 bottom right, 55 top left.

**Executive Editor** Trevor Davies
**Editor** Emma Pattison
**Executive Art Editor** Leigh Jones
**Designer** Ginny Zeal
**Production Manager** Ian Paton
**Illustrator** Mark Campion